Introduction to Geometric Algebra Computing

Introduction to Geometric Algebra Computing

Dietmar Hildenbrand

CRC Press
Taylor & Francis Group
Boca Raton London New York

CRC Press is an imprint of the
Taylor & Francis Group, an **informa** business

CRC Press
Taylor & Francis Group
6000 Broken Sound Parkway NW, Suite 300
Boca Raton, FL 33487-2742

© 2019 by Taylor & Francis Group, LLC
CRC Press is an imprint of Taylor & Francis Group, an Informa business

No claim to original U.S. Government works

Printed on acid-free paper
Version Date: 20180705

International Standard Book Number-13: 978-1-498-74838-4 (Hardback)

Visit the Taylor & Francis Web site at
http://www.taylorandfrancis.com

and the CRC Press Web site at
http://www.crcpress.com

To my beloved wife Carola
who died in March 2018
for her faith, hope and love.

Contents

SECTION II **Mathematical Foundations**

CHAPTER 4 ▪ Mathematical Basics and 2D Euclidean Geometric Algebra 45

Foreword

Dietmar Hildenbrand's new book *Introduction to Geometric Algebra Comput-ing* in my view fills an important gap in Clifford's geometric algebra literature. Most books up till now are written for the level of college students or above, pre-supposing substantial levels of linear-, vector- and matrix algebra, complex numbers, calculus, etc. The current book though starts with the elementary two-dimensional Euclidean geometry as I remember it from my middle school days. Therefore I expect that this book can be of interest beginning at the middle school level, through high school and college education. Furthermore engineers, who look for a beginner's introduction to geometric algebra are also at the right address. Because the textbook pedagogically begins by defining all geometric objects, the measuring of distance and angle, and elementary ge-ometric transformations, after having shown how to work interactively on the universal (can be implemented on all computing platforms) GAALOPScript. The latter permits the reader to immediately try and vary every new item, definition, entity, computation, etc. This is greatly assisted by the extensive set of GAALOPScript listings. In this sense the textbook is fully (inter)activ; the reader does not have to persevere through extensive theoretical founda-tions, before reaching the first motivating application, or explorative problem. At his own pace the reader can try the listed code of a section, and study the accompanying text for deeper understanding.

In geometric algebra, and especially so in conformal geometric algebra (CGA) of the two-dimensional plane, here appropriately named compass-ruler algebra (CRA), all expressions formed by elementary algebraic rules for ad-dition and multiplication are geometrically meaningful and useful. This is beautifully shown in Chapter 3 in the sections on difference and sum of two points, with respective interpretations as mid-line and circle (with the two points as poles), respectively. The reader can thus be assured, that he will not read any vain material, that later is of no or little use, and could safely be skipped.

In Section II, the mathematical basics of Chapter 4 will certainly help the reader to close the gap with college level GA textbooks. A notable aspect of Chapter 5 on CRA is that it elegantly shows how the notions of circle and line become unified in CRA, the line, so to speak, being a circle through infinity. Chapter 6, as elementary as it is, shows how simple products in CRA produce full geometric intersections of all geometric objects defined so far. Note, that this fully replaces the solution of linear systems of equations for the sets of intersection; the very same happens in CGAs for higher dimensional Euclidean (and non-Euclidean) spaces. True to the philosophy, that no computation is done in vain in CRA, Chapter 7 shows that even products of non-intersecting entities provide relative geometric information, such as distance. Elementary geometric compass and ruler constructions provide for geometric transformations (beginning with reflections at points, circles and lines), for which e.g. the mirror line encloses the object to be transformed like the two slices of a sandwich enclose a delicious slice of Iberian ham. This applies to any object to be transformed, and to any transformer (mirror line, mirror circle, mirror point, or their products, in short versors).

Section III focuses on applications, and given the worldwide enthusiasm of many teenagers for computer games and humanoid robots, the head start into robot kinematics (Chapter 9), is a most fortuitous choice by the author especially as he can base this introduction on his own highly original research and development in this field. Chapters 10 to 12 show how a conventional edge detection can benefit from being transcribed into CGA, successively simplified and optimized, and introduces two-dimensional visibility treatment in CGA (extendable to higher dimensions). The result is much easier to understand geometrically, and the elimination of all unnecessary computations and the natural parallel structure of GA computations, makes it ideal for high speed real time applications, including optimization of configurable hardware architecture. Section III closes with application to robotic snake control (unified treatment of kinematics and singular positions) and a brief view on CGA of three Euclidean dimensions, which at this level the reader will enjoy to explore interactively as well.

Section IV makes further proposals for the use of geometric algebra in school, including for high school level an introduction to two-dimensional space time physics, which remarkably relies on the same underlying algebra as CRA. Dietmar Hildenbrand has experience presenting CRA to pupils of all ages, and his colleague Martin E. Horn repeatedly taught high school students special relativity based on space-time algebra.

Altogether, I can only congratulate the author for the daring simplicity of his novel educational approach taken in this book, consequently combined with hands on computer based exploration. Without noticing, the active reader will thus educate himself in elementary geometric algebra algorithm development, geometrically intuitive, highly comprehensible, and fully optimized (remembering the meaning of GAALOP as Geometric Algebra Algorithm Optimizer).

Tokyo, 28th December 2017,
Eckhard Hitzer

Preface

Geometric Algebra is a very powerful mathematical system for an easy and intuitive treatment of geometry, but the community working with it is still very small. The main goal of this book is to close this gap with an introduction of Geometric Algebra from an engineering/computing perspective. The intended audience is students, engineers and researchers interested in learning Geometric Algebra and how to compute with it.

When I started to work with Geometric Algebra in 2003, I was immediately very impressed with how easy it is to develop 3D algorithms dealing with geometric objects and operations based on Geometric Algebra. I was very happy to use a tool providing me an immediate visual result for mathematical expressions: CLUCalc from Christian Perwass. I am still developing most of my Geometric Algebra algorithms with CLUCalc and, in my opinion, CLUCalc is still today the best tool in order to learn how to use Geometric Algebra for 3D applications.

In 2004, when I organized and presented a Geometric Algebra tutorial at the Eurographics conference in Grenoble together with Christian Perwass, Daniel Fontijne and Leo Dorst, the feedback was positive and negative at the same time. On one hand, many people were happy about the expressiveness of Geometric Algebra, but on the other hand it was clear for everybody that implementations of computer graphics applications were not competitive in terms of runtime performance. I realized that improving the runtime performance of Geometric Algebra will be the key to convince engineers to use Geometric Algebra in their applications.

At that time, nobody really expected that it could be possible for implementations of Geometric Algebra algorithms to be faster than the conventional implementation. But, in 2006, we were happy to present even two different implementations proving exactly that for a computer animation application (the movement of the arm of a virtual character). Our approach was very specific for this proof-of-concept application. This is why our next goal was a general system making it possible for almost every engineer to include Geometric Algebra in his/her application. And, the description of Geometric Algebra algorithms should be as much as possible similar to how CLUCalc is doing that. Now, we are happy to provide the *GAALOP* [1] *(GEOMETRIC ALGEBRA ALGORITHMS OPTIMIZER)* precompiler for the integration of Geometric

---

[1] GAALOP is henceforth written in capital letters

Algebra into standard programming languages such as C++, OpenCL, CUDA and C++ AMP. The integration is done based on *GAALOPScript*, which is very much inspired by the CLUCalc scripting language. This technology is described in my book *Foundations of Geometric Algebra Computing* [26] from 2013. Since 2015 this technology is part of the ecosystem of the HSA Foundation of more than 40 companies dealing with new heterogeneous computing architectures.

Today, we indeed have this Geometric Algebra Computing technology available for easy to develop, geometrically intuitive, robust and fast engineering applications, but there is still only a small number of people who know it. Exactly at this point this book comes into place. This book is intended to give a rapid introduction to the computing with Geometric Algebra and its power for geometric modeling. From the geometric objects' point of view it focuses on the most basic ones, namely points, lines and circles. We call this algebra *Compass Ruler Algebra*, since you are able to handle it comparable to working with compass and ruler. It offers the possibility to compute with these geometric objects, their geometric operations and transformations in a very intuitive way. While focusing on 2D it is easily expandable to 3D computations as used in many books dealing with the very popular Conformal Geometric Algebra in engineering applications such as computer graphics, computer vision and robotics. Throughout this book, we use *GAALOPScript*, the input language of GAALOP, in order to describe and visualize the algorithms and in order to generate C/C++ or LaTeX code helping us to look behind the scene. This book follows a top-down approach. Focusing first on how to use Geometric Algebra, it is up to the reader how much he/she would like to go into the details.

Recently, we could celebrate the 50th anniversary of the book *Space-Time Algebra*[2] of David Hestenes. Published in 1966, it was the starting point for his very fruitful Geometric Algebra research. Especially important for this book is his work on Conformal Geometric Algebra. Interestingly, the Space-Time Algebra of David Hestenes and the Compass Ruler Algebra as a specific Conformal Geometric Algebra (both treated in this book), have a similar algebraic structure. Maybe some teachers will use this book for teaching themselves and for introducing it already in school?

I really do hope that this book can support the widespread use of Geometric Algebra as a mathematical tool for computing with geometry.

Dietmar Hildenbrand

---

[2] 1st edn. [15], 2nd edn. with a foreword by Anthony Lasenby[22]

Acknowledgments

I would like to thank my former student Christian Steinmetz for his tremendous support of this book. He developed GAALOP further as used in this book during his bachelor's and master's thesis and he is still an active developer of GAALOP which is now an open source software project. I am very grateful for many improvements of GAALOP that he made while writing this book.

Special thanks to Prof. Petr Vasik, J. Hrdina and A. Navrat from Brno University of Technology for their support of a nice chapter about the control of a snake robot.

Many thanks to

- Prof. Yu Zhaoyuan and Dr. Werner Benger for the joint work dealing with the inner products of geometric objects as well as for their very helpful reviews of the book,

- Prof. Eduardo Bayro-Corrochano and his group for their nice application dealing with the detection of circles and lines in images, which is a very good example for the runtime considerations in this book,

- Dr. Silvia Franchini, Prof. A. Gentile, Prof. G. Vassallo and Prof. S. Vitabile from the University of Palermo for the good cooperation regarding a new co-processor design for Geometric Algebra,

- Senior Associate Prof. Eckhard Hitzer for many fruitful discussions, for his papers [37] and [39] as inspirations for this book and for his big effort and enthusiasm for the promotion of Geometric Algebra (and the computing with it) as the president of the International Advisory Board of the International Conference on Clifford Algebras and Their Applications in Mathematical Physics (ICCA) and as the main organizer of many workshops and conferences in the field,

- Janina Osti for proofreading an early version of the book from the perspective of a first semester student.

Since we recently could celebrate **the 50th anniversary of the book** *Space-Time Algebra* **of David Hestenes**, I am grateful for the support of a chapter dealing with this algebra

- Dr. Martin E. Horn for his support in describing his way of using Space-Time-Algebra in school,

- Mariusz Klimek for his simulation application based on Space-Time Algebra,

- my former student Patrick Charrier for his adaptation of the GAALOP Precompiler, which he developed in his master's thesis, to this Space-Time Algebra application.

Introduction

CONTENTS

This book serves as an introduction to Geometric Algebra from a computing/engineering perspective. Its goal is to develop an appetite for delving into more. It pushes the reader into the cold water of swimming with Geometric Algebra right away, showing how to do things and what can be done, without the often burdensome overhead of rigid mathematical definitions.

1.1 Geometric Algebra

Geometric Algebra is a mathematical framework that makes it easy to describe geometric concepts and operations. It allows us to develop algorithms fast and in an intuitive way.

Geometric Algebra is based on the work of the German high school teacher Hermann Grassmann and his vision of a general mathematical language for geometry. His very fundamental work, called *Ausdehnungslehre*[14], was little noted in his time. Today, however, Grassmann is more and more respected as one of the most important mathematicians of the 19th century. William Clifford [5] combined Grassmann's exterior algebra and Hamilton's quaternions in what we call *Clifford algebra* or *Geometric Algebra*[1].

---

[1]David Hestenes writes in his article [23] about the genesis of Geometric Algebra: Even today mathematicians typically typecast Clifford Algebra as the algebra of a quadratic form, with no awareness of its grander role in unifying geometry and algebra as envisaged by Clifford himself when he named it Geometric Algebra. It has been my privilege to pick up where Clifford left off to serve, so to speak, as principal architect of Geometric Algebra and Calculus as a comprehensive mathematical language for physics, engineering and computer science.

Pioneering work has been done by David Hestenes, 50 years ago. His book *Space-Time Algebra* [15] was the starting point for his development of Geometric Algebra into a unified mathematical language for physics, engineering and mathematics [16, 24] [20]. Especially interesting for this book is his work on Conformal Geometric Algebra (CGA) [17] [50]: the *Compass Ruler Algebra* (CRA), mainly treated in this book, is simply the Conformal Geometric Algebra in 2D.

The main advantage of Geometric Algebra is its easy and intuitive treatment of geometry. This is why the focus of this book is on the introduction of Geometric Algebra based on the computing with the most basic geometric objects, namely points, lines and circles. While we are computing in 2D space, the underlying algebra is the 4D Compass Ruler Algebra with a close link between algebra and the geometry of these basic geometric objects. While focusing on 2D, it is easily expandable to 3D computations as used, for instance, in the books [26], [57], [1] and [8].

1.2 Geometric Algebra Computing

Especially since the introduction of Conformal Geometric Algebra there has been an increasing interest in using Geometric Algebra in engineering. The use of Geometric Algebra in engineering applications relies heavily on the availability of an appropriate computing technology. The main problem of *Geometric Algebra Computing* is the exponential growth of data and computations compared to linear algebra, since the **multivector**[2] of an n-dimensional Geometric Algebra is 2^n-dimensional. For the 5-dimensional Conformal Geometric Algebra, the multivector is already 32-dimensional.

An approach to overcome the runtime limitations of Geometric Algebra has been through optimized software solutions. Tools have been developed for high-performance implementations of Geometric Algebra algorithms such as the C++ software library generator Gaigen 2 from Daniel Fontijne and Leo Dorst of the University of Amsterdam [11], GMac from Ahmad Hosney Awad Eid of Suez Canal University [10], the Versor library [6] from Pablo Colapinto, the C++ expression template library Gaalet [63] from Florian Seybold of the University of Stuttgart, and our GAALOP compiler [29], which can also be used as a precompiler for languages such as C/C++, CUDA, OpenCL and C++ AMP. The big potential of optimizations of Geometric Algebra algorithms can be very well demonstrated with the inverse kinematics algorithm of [30] [25], which was in 2006 the first Geometric Algebra application that was faster than the standard implementation.

The book *Foundations of Geometric Algebra Computing* [26] defines Geometric Algebra Computing as the geometrically intuitive development of algorithms using Geometric Algebra with a focus on their efficient implementation. It describes Geometric Algebra Computing in a very fundamental way, since

---

[2]The main algebraic element of Geometric Algebra (please refer to Chapt. 2)

it breaks down the computing of Geometric Algebra algorithms to the most basic arithmetic operations.

This book on hand makes use of *GAALOP* [29] as a free and easy to handle tool in order to compute and visualize with Geometric Algebra. The book is suitable as a starting point for the understanding of Geometric Algebra for everybody interested in it as a new powerful mathematical system, especially for students, engineers and researchers in engineering, computer science and mathematics.

1.3 OUTLINE

This book is organized in the following sections:

I Tutorial

II Mathematical Foundations

III Applications

IV Geometric Algebra at School

SECTION I is a tutorial on how to work with Geometric Algebra, especially with Compass Ruler Algebra and its geometric objects, namely circles, lines and point pairs. SECTION II is for readers, now interested in the mathematical background of what they did in SECTION I. Readers, more interested in applications, are able to directly switch to SECTION III with applications in the areas of robotics, computer vision and computer graphics. SECTION IV gives some considerations about using Geometric Algebra already at school and about *Space-Time Algebra* in honor of the work of David Hestenes and especially to the 50th anniversary of his book about this algebra.

1.3.1 SECTION I : Tutorial

Chapt. 2 presents Compass Ruler Algebra in a nutshell as an algebra of circles, lines and point pairs. It summarizes the algebraic expressions needed for the tutorial in Chapt. 3 in order to describe the geometric objects and their intersections, the angles and distances between them as well as their reflections, rotations and translations.

Chapt. 3 is an easy to understand tutorial for what we learned in Chapt. 2. It makes use of GAALOP (see [26]) as a free and easy to handle tool in order to compute and visualize with Compass Ruler Algebra. This chapter is equipped with simple examples. We will see, for instance, how easy it is to deal with bisectors or with the circumcircle of a triangle. This chapter is written in a tutorial-like style in order to encourage the reader to gain his/her own experience in developing algorithms based on Compass Ruler Algebra.

1.3.2 SECTION II : Mathematical Foundations

Chapt. 4 presents some mathematical background on Geometric Algebra in general, with a focus on the 2D Euclidean Geometric Algebra. It introduces the basic algebraic elements as well as the main products, namely the inner, outer and geometric product in more detail.

Chapt. 5 introduces the Compass Ruler Algebra as a Geometric Algebra with which you are able to compute in a way similar to working with compass and ruler. We describe the algebraic structure and the representations of geometric objects. For Compass Ruler Algebra computations, we use the GAALOP software package.

In Geometric Algebra, the outer product can be used for the intersection of geometric objects. In Compass Ruler Algebra, the intersection of circles and lines are so-called point pairs, which are described in **Chapt. 6** in some detail.

Chapt. 7 focuses on computations based on the inner product describing distances and angles between the basic geometric objects of Compass Ruler Algebra.

Chapt. 8 describes transformations in Compass Ruler Algebra. Based on the geometric product, reflections of the basic geometric objects can be expressed. Transformations such as rotations and translations of the basic geometric objects can be expressed as consecutively executed reflections or based on specific operators called rotors and translators.

1.3.3 SECTION III : Applications

SECTION III starts with applications from robotics, computer vision and computer graphics using Gaalop

> **Chapt. 9** deals with a robot kinematics application of moving a robot to a target position.

> **Chapt. 10** presents an application dealing with the detection of circles and lines in images.

> **Chapt. 11** describes an application in 2D which is easily expandable to 3D: computing the visibiliy of bounded spheres related to a view cone.

> **Chapt. 12** presents some considerations about runtime-performance of applications using GAALOP.

Chapt. 13 presents the fitting of geometric objects into point clouds based on Compass Ruler Algebra. Since lines as well as circles have the same algebraic structure (both are represented by vectors), it is easy to make an approach in order to fit the best geometric object into a set of points whether it is a circle or a line.

Chapt. 14 treats robot kinematics in a mathematically advanced manner: the control of a snake robot is presented based on differential kinematics.

While the Geometric Algebra introduction of this book is based on computations in 2D space, **Chapt. 15** is dedicated to their expansion to 3D. Based on these explanations, it will be easy for the reader to follow the literature dealing with Geometric Algebra applications in 3D.

1.3.4   SECTION IV : Geometric Algebra at School

SECTION IV is added in honor of the work of David Hestenes in education and the 50th anniversary of his book about *Space-Time Algebra*.

Chapt. 16 reviews some thoughts about the potential use of Geometric Algebra in schools. It tries to answer the question whether GAALOP can be the basis for an appropriate tool for mathematical education based on Geometric Algebra.

Chapt. 17 handles Space-Time Algebra on one hand based on easy examples that can be treated already in schools. On the other hand, a simple physical simulation based on a moving particle is shown.

I

Tutorial

Compass Ruler Algebra in a Nutshell

CONTENTS

The two chapters of SECTION I cover a tutorial on how to work with Compass Ruler Algebra[1]. This chapter summarizes the algebraic expressions needed for the examples of Chapt. 3.

TABLE 2.1   Notations of Compass Ruler Algebra.

Notation	Meaning	Details in chapter
AB	geometric product of A and B	4
$A \wedge B$	outer product of A and B	4
$A \cdot B$	inner product of A and B	4
A^{-1}	inverse of A	4
A^*	dual of A	4
$\tilde{A}$	reverse of A	4
e_1, e_2	2D basis vectors	4
$i = e_1 \wedge e_2$	imaginary unit	4
e_0	origin	5
e_∞	infinity	5

Table 2.1 summarizes the most important notations of Compass Ruler Algebra. The three main products of Geometric Algebra are the geometric, the outer and the inner product. Please notice that for the geometric product no specific symbol is used. Important operations of Geometric Algebra are the

---

[1]Simply the Conformal Geometric Algebra [17] [50] in 2D as a Geometric Algebra of circles, lines and point pairs.

TABLE 2.2 The 16 basis blades of the Compass Ruler Algebra (to be identified by their indices).

Index	Blade
0	1
1	e_1
2	e_2
3	e_∞
4	e_0
5	$e_1 \wedge e_2$
6	$e_1 \wedge e_\infty$
7	$e_1 \wedge e_0$
8	$e_2 \wedge e_\infty$
9	$e_2 \wedge e_0$
10	$e_\infty \wedge e_0$
11	$e_1 \wedge e_2 \wedge e_\infty$
12	$e_1 \wedge e_2 \wedge e_0$
13	$e_1 \wedge e_\infty \wedge e_0$
14	$e_2 \wedge e_\infty \wedge e_0$
15	$e_1 \wedge e_2 \wedge e_\infty \wedge e_0$

inverse, dual and reverse operations (see Chapt. 4 for details). e_1 and e_2 are the 2D basis vectors in x- and y-direction. The imaginary unit i with the property $i^2 = -1$ can be identified as the outer product of the two basis vectors. $e_1 \wedge e_2$ is one example for a basis **blade** consisting of combinations of outer products of e_1 and e_2 and two additional basis vectors of Compass Ruler Algebra, e_0 and e_∞, according to Table 2.2. Linear combinations of these basis blades are called **multivectors**, which are the main algebraic elements of Geometric Algebra. Please refer to Chapt. 5 for details about the algebraic structure of Compass Ruler Algebra.

In the next sections, we will see how

- geometric objects and their intersections

- angles and distances between geometric objects

- transformations of geometric objects

can be expressed easily with the help of algebraic expressions.

2.1 GEOMETRIC OBJECTS

Table 2.3 shows a list of the basic geometric objects of the Compass Ruler Algebra, namely points, circles, lines and point pairs.

TABLE 2.3 The representations of the geometric objects of the Compass Ruler Algebra.

Entity	standard representation	dual representation
Point	$P = \mathbf{x} + \frac{1}{2}\mathbf{x}^2 e_\infty + e_0$	
Circle	$C = P - \frac{1}{2}r^2 e_\infty$	$C^* = P_1 \wedge P_2 \wedge P_3$
Line	$L = \mathbf{n} + d e_\infty$	$L^* = P_1 \wedge P_2 \wedge e_\infty$
Point pair	$P_p = C_1 \wedge C_2$	$P_p^* = P_1 \wedge P_2$

These entities have two algebraic representations, the standard and the dual representation. These representations are duals of each other (a superscript asterisk denotes the dualization operator). A 2D point with coefficients x_1, x_2 and basis vectors e_1, e_2

$$\mathbf{x} = x_1 e_1 + x_2 e_2 \qquad (2.1)$$

is embedded in the 4D Compass Ruler Algebra as point

$$P = \mathbf{x} + \frac{1}{2}\mathbf{x}^2 e_\infty + e_0 \qquad (2.2)$$

with the two additional basis vectors e_∞, e_0 (with the geometric meaning of infinity and origin) and

$$\mathbf{x}^2 = x_1^2 + x_2^2 \qquad (2.3)$$

being the scalar product of $\mathbf{x}$.

$\mathbf{x}$ and $\mathbf{n}$ in Table 2.3 are in bold type to indicate that they represent 2D entities obtained by linear combinations of the 2D basis vectors e_1 and e_2. L represents a line with normal vector $\mathbf{n}$ and distance d to the origin. The $\{C_i\}$ represent different circles. The outer product "$\wedge$" indicates the construction of a geometric object with the help of points $\{P_i\}$ that lie on it. A circle, for instance, is defined by three points $(P_1 \wedge P_2 \wedge P_3)$ on this circle. Another meaning of the outer product is the intersection of geometric entities. A point pair is defined by the intersection of two circles $C_1 \wedge C_2$. Please refer to Chapt. 5 for more details about the geometric objects of Compass Ruler Algebra.

2.2 ANGLES AND DISTANCES

The inner product of these geometric objects describes distances and angles between them as summarized in Table 2.4. The inner product $L_1 \cdot L_2$ of two lines L_1 and L_2, for instance, describes the angle between these lines, while the inner product between other geometric objects describes the Euclidean distance or some kind of distance measure between them. Please refer to Chapt. 7 for more details.

TABLE 2.4 Geometric meaning of the inner product of lines, circles and points.

$A \cdot B$	B **Line**	B **Circle**	B **Point**
A **Line**	Angle between lines	Euclidean distance from center	Euclidean distance
A **Circle**	Euclidean distance from center	Distance measure	Distance measure
A **Point**	Euclidean distance	Distance measure	Distance

TABLE 2.5 The description of transformations of a geometric object o in Compass Ruler Algebra (please note that LoL means the geometric product of L, o and L).

Transformation	operator	usage
Reflection	$L = \mathbf{n} + d e_\infty$	$o_L = -LoL$
Rotation	$R = \cos\left(\frac{\phi}{2}\right) - i \sin\left(\frac{\phi}{2}\right)$	$o_R = Ro\tilde{R}$
Translation	$T = 1 - \frac{1}{2}\mathbf{t}e_\infty$	$o_T = To\tilde{T}$

2.3 TRANSFORMATIONS

Transformations of a geometric object o can be easily described within Compass Ruler Algebra according to Table 2.5. The reflection, for instance, of a circle C at a line L can be computed based on the (geometric) product $-LCL$ (please remember that the geometric product in Geometric Algebra is written without a specific product symbol). Rotations or translations can be described based on algebraic expressions called rotors R and translators T. Using the rotor

$$R = \cos\left(\frac{\phi}{2}\right) - i \sin\left(\frac{\phi}{2}\right), \tag{2.4}$$

the rotated object o_R can be computed based on the geometric product $Ro\tilde{R}$

$$o_R = Ro\tilde{R} \tag{2.5}$$

with $\tilde{R}$ being the *reverse* of R (see Sect. 4.6). A translated object o_T can be computed based on the translator

$$T = 1 - \frac{1}{2}\mathbf{t}e_\infty \tag{2.6}$$

with $\mathbf{t}$ being the 2D translation vector $t_1 e_1 + t_2 e_2$ as

$$o_T = To\tilde{T}. \tag{2.7}$$

Please refer to Chapt. 8 for more details.

CHAPTER 3

GAALOP Tutorial for Compass Ruler Algebra

CONTENTS

This chapter is an easy to understand tutorial for what we learned in Chapt. 2. It makes use of GAALOP (see [26]) as a free and easy to handle tool in order to compute and visualize with Compass Ruler Algebra based on simple examples. We will see, for instance, how easy it is to deal with bisectors or with the circumcircle of a triangle. This chapter is written in a tutorial-like style in order to encourage the reader to gain his/her own experience in developing algorithms based on Compass Ruler Algebra. Mathematical details will follow in subsequent chapters.

3.1 GAALOP AND GAALOPSCRIPT

We use the GAALOP software to compute with Compass Ruler Algebra and to visualize the results of these computations. GAALOP [29] can be downloaded free of charge from *http://www.gaalop.de*.

We recommend also to install Maxima [53] in order to be able to use the complete optimization potential of GAALOP. Fig. 3.1 shows how GAALOP

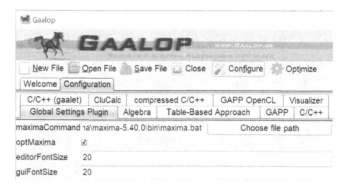

FIGURE 3.1 Global Setting Plugin for the configuration of Maxima (as well as font sizes).

has to be configured for the use of Maxima[1]. In the Global Setting Plugin the path of the file *maxima.bat* of the Maxima installation has to be chosen and *optMaxima* has to be activated.

FIGURE 3.2 Configuration of GAALOP for visualizations based on Compass Ruler Algebra.

---

[1]Maxima is also used in this book for function diagrams and some symbolic computations.

The screenshot in Fig. 3.2 shows how GAALOP should be configured for the visualizations of this tutorial. Please select the "cr4d - compass-ruler" as "Algebra to use" and the "Visual Code Inserter 2d" for 2d visualizations to be performed by "Vis2d", the "CodeGenerator" to be selected for our purpose. Please also select the default "Table-Based Approach" for "Optimization".

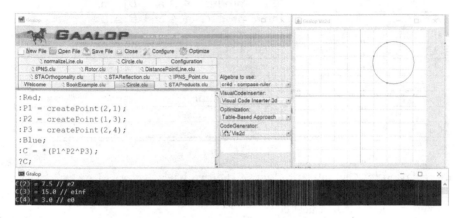

FIGURE 3.3 Screenshot of the editor, the visualization and the output window of GAALOP.

The screenshot in Fig. 3.3 shows the three windows of GAALOP. The editor window is responsible for the administration and for the editing of *GAALOPScripts* (the input language of GAALOP). GAALOPScript is essentially a subset of CLUScript, the input language of CLUCalc[2] [58], adapted to 2D computations. Aside is the CodeGenerator/visualization window. If Vis2d is selected for 2D visualizations of GAALOPScripts, the output window (at the bottom) is able to show numeric values of multivectors[3]. While the visualization window is mainly used in SECTION I (Tutorial), the LaTeX code generator is mainly used in SECTION II (Mathematical Foundations) and the C/C++ code generator in SECTION III (Applications). These code generators describe multivectors with their coefficients based on the indices of Table 2.2. The tutorial is based on some simple examples, highlighting aspects of Compass Ruler Algebra. These examples are meant as starting points for your own experiments. We hope that they will inspire you to make your own changes and gain your own experience with Compass Ruler Algebra.

Table 3.1 summarizes the most important notations of GAALOPScript for Compass Ruler Algebra. Please notice that while for the geometric product no specific symbol is used, in GAALOP "*" is needed as symbol. The inverse of A is expressed as "1/A". The operators for the dual and the reverse of A are written in front of A. The imaginary unit i is expressed as "e12", an

---

[2]CLUCalc is a free tool which can be used for 3D visualizations of Geometric Algebra (we will use it in Chapt. 15).

[3]Please refer to Chapt. 2

TABLE 3.1 Notations of Compass Ruler Algebra in GAALOPScript

Compass Ruler Algebra	Meaning	GAALOPScript
AB	geometric product	A*B
$A \wedge B$	outer product of A and B	A∧B
$A \cdot B$	inner product of A and B	A.B
A^{-1}	inverse of A	1/A
A^*	dual of A	*A
$\tilde{A}$	reverse of A	~A
e_1, e_2	2D basis vectors	e1, e2
i	imaginary unit	e12
e_0	origin	e0
e_∞	infinity	einf

abbreviation for the outer product of the two basis vectors $e1$ and $e2$ (please refer to Chapt. 2). The other GAALOP notations will be explained in the following sections when needed.

3.2   GEOMETRIC OBJECTS

We learned how to describe geometric objects of Compass Ruler Algebra in Sect. 2.1 on page 10. Table 3.2 shows, how to express them in GAALOPScript (dependent on the standard and dual representation).

TABLE 3.2   The two representations of the geometric objects of the Compass Ruler Algebra expressed in GAALOPScript.

Entity	standard representation	GAALOPScript
Point	$P = \mathbf{x} + \frac{1}{2}\mathbf{x}^2 e_\infty + e_0$	P=createPoint(x1,x2)
Circle	$C = P - \frac{1}{2}r^2 e_\infty$	C=P-0.5*r*r*einf
Line	$L = \mathbf{n} + d e_\infty = n_1 e_1 + n_2 e_2 + d e_\infty$	L= n1*e1+n2*e2+d*einf
Point pair	$P_p = C_1 \wedge C_2$	Pp = C1∧C2
	dual representation	
Circle	$C^* = P_1 \wedge P_2 \wedge P_3$	C = *(P1∧P2∧P3)
Line	$L^* = P_1 \wedge P_2 \wedge e_\infty$	L=*(P1∧P2∧einf)
Point pair	$P_p^* = P_1 \wedge P_2$	Pp=*(P1∧P2)

3.2.1 Point

The GaalopScript[4] according to Listing 3.1 shows two possible ways of defining points based on GAALOP.

Listing 3.1 *Point.clu*: GAALOPScript for two different definitions of points.

```
1  x1 = 2;
2  x2 = 1;
3  P1 = x1*e1 + x2*e2 + 0.5*(x1*x1 + x2*x2)*einf + e0;
4  P2 = createPoint(x1,x2);
5
6  // visualize the points
7  :P1;
8  :P2;
9
10 // numerical output of the points
11 ?P1;
12 ?P2;
```

$P1$ is defined explicitly based on Eq. (2.2) while $P2$ is defined based on the predefined macro *createPoint()*, both describing the same 2D point (2,1). "//" is used in GAALOPScripts for comments (as also usual in C/C++).

A **leading colon** means the indication of multivectors which should be visualized. Since both points have the same coordinates, only one point at the location (2,1) is drawn by GAALOP Vis2d according to Fig. 3.4.

FIGURE 3.4 Visualization of *Point.clu*.

A **leading question** mark indicates that the numerical values of the corresponding multivector should be shown in the output window. Since $P1$ and

---

[4]Please notice that all the GAALOPScripts of this book can be downloaded from *http://www.gaalop.de*.

$P2$ describe the same point, their e_1, e_2, e_∞, e_0-components are the same. This can be seen in the result of the output window according to Listing 3.2 where the points $P1$ and $P2$ have the same numerical values.

Listing 3.2 numerical output of *Point.clu*: Two different definitions of points show the same numerical results.

```
1  P1(1) = 2.0 // e1
2  P1(2) = 1.0 // e2
3  P1(3) = 2.5 // einf
4  P1(4) = 1.0 // e0
5  P2(1) = 2.0 // e1
6  P2(2) = 1.0 // e2
7  P2(3) = 2.5 // einf
8  P2(4) = 1.0 // e0
```

In general, the multivectors are indicated based on their non-zero components. $P1$ and $P2$ need only the components with indices 1, 2, 3 and 4 of the 16 basis blades according to Table 2.2. The comments at the end of each line show the names of these basis blades, namely e1, e2, einf (for e_∞) and e0.

3.2.2 Circle

The GAALOPScript of Listing 3.3 is our first script with definitions of colors. With the help of the Vis2d component, it is transformed into the visualization of Fig. 3.5.

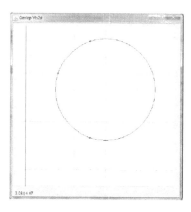

FIGURE 3.5 Visualization of *Circle1.clu*: a circle based on the outer product of three points.

First of all, three points with the 2D coordinates $(2, 1)$, $(1, 3)$ and $(2, 4)$ are transformed into 4D coordinates of the Compass Ruler Algebra and visualized in red. Then the circle C is computed based on the outer product of these three points, transformed into the IPNS representation via the dualization operator and visualized in blue. Note: this way the circumcircle of a triangle can be computed very easily.

Listing 3.3 *Circle1.clu*: Script for the visualization of a circle based on the outer product of three points.

```
1  : Red ;
2  : P1 = createPoint (2 ,1);
3  : P2 = createPoint (1 ,3);
4  : P3 = createPoint (2 ,4);
5  : Blue ;
6  : C = *( P1 ^ P2 ^ P3 );
7  ? C ;
```

The numerical values of the circle C are computed according to Listing 3.4.

Listing 3.4 Output result of the GAALOPScript *Circle1.clu*: the numerical values of the circle C.

```
1  C (1) = 7.5 // e1
2  C (2) = 7.5 // e2
3  C (3) = 15.0 // einf
4  C (4) = 3.0 // e0
```

Note: this circle is not normalized in the sense that its e_0-component is 1. If we are interested in the 2D location of its center point, we have to divide all components by 3. Then we get the correct 2D position (2.5, 2.5). Please refer to Sect. 5.6 for normalized objects.

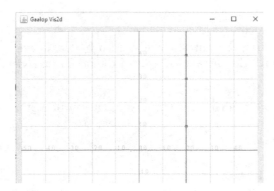

FIGURE 3.6 Visualization of *Circle2.clu*: a line based on the outer product of three co-linear points.

What happens, if the points are co-linear? Changing the GAALOPScript in order to have all the points on one line[5],

Listing 3.5 *Circle2.clu*: Script for the visualization of a line based on the outer product of three co-linear points.

```
1  : Red ;
2  : P1  =  createPoint (2 ,1) ;
3  : P2  =  createPoint (2 ,3) ;
4  : P3  =  createPoint (2 ,4) ;
5  : Blue ;
6  : C  =  *( P1 ^ P2 ^ P3 ) ;
7  ? C ;
```

the result is just this line (see Fig. 3.6). Looking at the output window according to Listing 3.6,

Listing 3.6 Output result of the changed GAALOPScript *Circle2.clu*: the numerical values of the specific circle C with infinite radius (a line).

```
1  C (1)  =  3.0  //  e1
2  C (3)  =  6.0  //  einf
```

we see that the e_0-component of C is missing. This indicates algebraically that the result is a line. The line is not normalized. Its normal vector $n = 3e_1$ has a length of 3. Dividing by 3 results in the expression

$$C = e_1 + 2e_\infty \tag{3.1}$$

which is a line with normal vector e_1 and a distance of 2 to the origin.

3.2.3 Line

The following listing computes a line defined by the normal vector n in the direction (1,1) and the distance $d = 2$ to the origin and visualizes it in Fig. 3.7.

Listing 3.7 *Line.clu*: Script for the visualization of a line.

```
1  n1  =  sqrt (2) /2 ;
2  n2  =  sqrt (2) /2 ;
3
4  n  =  n1 * e1  +  n2 * e2 ;
5  d  =  2 ;
6  : Line  =  n  +  d * einf ;
```

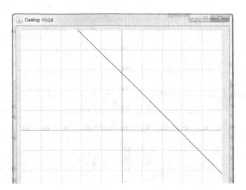

FIGURE 3.7 Visualization of *Line.clu*: a line based on the normal vector $\frac{1}{2}\sqrt{2} * (1, 1)$ and a distance of 2 to the origin.

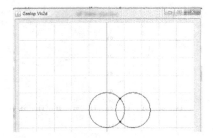

FIGURE 3.8 Visualization of *CircleCircleCut.clu*: a point pair as the intersection of two circles.

3.2.4 Point pair

The following listing computes a point pair PP based on the intersection of two circles $C1$ and $C2$ and visualizes these geometric objects in Fig. 3.8.

Listing 3.8 *CircleCircleCut.clu*: Script for the visualization of a point pair as the intersection of two circles.

```
1  d = 1;
2  r1 = 1;
3  r2 = 1;
4
5  :C1 = e0-0.5*r1*r1*einf;
6  :C2 = createPoint(d,0)-0.5*r2*r2*einf;
```

---

[5]According to Listing 3.5

```
7 | :PP = C1^C2;
```

3.2.5 Perpendicular Bisector

In order to construct the perpendicular bisector of a line segment with compass and ruler, we draw two circles with the center at the boundary points and connect the two intersection points according to Fig. 3.9.

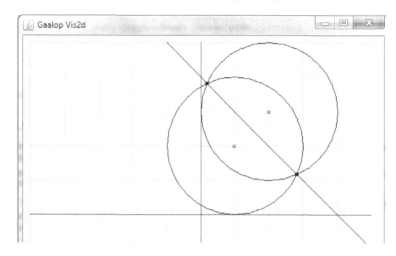

FIGURE 3.9 Visualization of the perpendicular bisector between the two (red) points.

How can we express this construction based on Compass Ruler Algebra?

Listing 3.9 *PerpendicularBisector.clu*: Computation of the perpendicular bisector of the section of line between the points P1 and P2 with the help of the intersection of two circles.

```
1  | P1 = createPoint(x1,y1);
2  | P2 = createPoint(x2,y2);
3  | // intersect two circles with center points P1 and P2
4  | // with the same, but arbitrary radius
5  | S1 = P1 - 0.5*r*r*einf;
6  | S2 = P2 - 0.5*r*r*einf;
7  | PPdual = *(S1^S2);
8  | // the line through the two points
9  | // of the resulting point pair
10 | Bisector = *(PPdual^einf);
```

This GAALOPScript computes, first of all, two points P1 and P2 with the symbolic 2D coordinates (x1,y1) and (x2,y2). Then, the circles S1 and S2 with center at the points P1 and P2 with radius r are computed. The intersection of the circles results in a point pair (see Table 2.3). Its dual *PPdual* describes the outer product of the two intersection points. Finally, the resulting bisector line is the dual of the outer product of these two points and e_∞ (see Table 2.3).

For the visualization of Fig. 3.9 the variables have to be equipped first with concrete input values (see Listing 3.10);

Listing 3.10 *PerpendicularBisector.clu*: concrete input values for Fig. 3.9.

```
1  x1 = 1;
2  y1 = 2;
3  x2 = 2;
4  y2 = 3;
5  r = 2;
```

as well the colors for the geometric objects to be drawn have to be defined at the end (see Listing 3.11).

Listing 3.11 *PerpendicularBisector.clu*: Visualization statements for Fig. 3.9 .

```
1  :Red;
2  :P1;
3  :P2;
4  :Black;
5  :S1;
6  :S2;
7  :Bisector;
```

The points are visualized in red and the circles and the bisector line in black.

3.2.6 The Difference of two Points

Listing 3.12 computes the difference of two points. This results in the line in the middle between the two points, which is visualized in Fig. 3.10.

Listing 3.12 *DifferencePointPoint.clu*: Script for the visualization of the difference of two points.

```
1  p1 = 1;
2  p2 = 2;
3  q1 = 0;
4  q2 = 1;
5
6  P = createPoint(p1,p2);
7  Q = createPoint(q1,q2);
8  Diff = P-Q;
9
```

FIGURE 3.10 Visualization of *DifferencePointPoint.clu*: the difference of two points.

```
10 | : Red ;
11 | : P ;
12 | : Q ;
13 | : Green ;
14 | : Diff ;
```

This means that the perpendicular bisector of a line segment can be easily computed based on the difference of its vertices. Is that true in arbitrary cases? We will show that in Sect. 16.3.

3.2.7 The Sum of Points

FIGURE 3.11 Visualization of *SumOfTwoPoints.clu*: the sum of two points.

Listing 3.13 computes the sum of two points resulting in their minimal enclosing circle, which is visualized in Fig. 3.11. The circle is dashed indicating that its radius is imaginary, which means the square of the radius is smaller than zero.

Listing 3.13 *SumOfTwoPoints.clu*: Script for the visualization of the sum of two points.

```
1  p1=1;
2  p2=1;
3  q1=3;
4  q2=2;
5
6  P = createPoint(p1,p2);
7  Q = createPoint(q1,q2);
8  ?C = P+Q;
9
10 :Red;
11 :P;
12 :Q;
13 :Black;
14 :C;
```

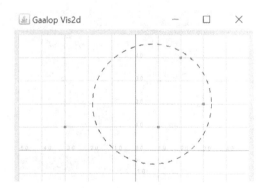

FIGURE 3.12 Visualization of *SumOfPoints.clu*: the sum of four points.

What happens when taking more than two points? Listing 3.14 computes the sum of four points resulting in some kind of approximation by a circle, which is visualized in Fig. 3.12.

Listing 3.14 *SumOfPoints.clu*: Script for the visualization of the sum of four points.

```
1  p1=1;
2  p2=1;
```

```
3   q1=3;
4   q2=2;
5   r1=2;
6   r2=4;
7   s1=-3;
8   s2=1;
9   P = createPoint(p1,p2);
10  Q = createPoint(q1,q2);
11  R = createPoint(r1,r2);
12  S = createPoint(s1,s2);
13  C = P+Q+R+S;
14
15  :Red;
16  :P;
17  :Q;
18  :R;
19  :S;
20  :Black;
21  :C;
```

We realize that the sum of points can be used for some kind of fitting a circle into a set of points.

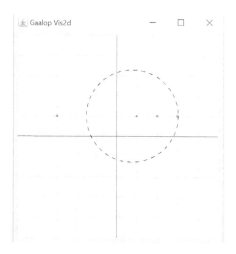

FIGURE 3.13 Visualization of *SumOfCOLinearPoints.clu*: the sum of four points lying on one straight line.

But, what happens if these points are on a straight line? Listing 3.15 computes the sum of four points on one line.

Listing 3.15 *SumOfCOLinearPoints.clu*: Script for the visualization of the sum of four co-linear points.

```
1  p1=1;
2  p2=1;
3  q1=3;
4  q2=1;
5  r1=2;
6  r2=1;
7  s1=-3;
8  s2=1;
9  P = createPoint(p1,p2);
10 Q = createPoint(q1,q2);
11 R = createPoint(r1,r2);
12 S = createPoint(s1,s2);
13 C = P+Q+R+S;
14
15 :Red;
16 :P;
17 :Q;
18 :R;
19 :S;
20 :Black;
21 :C;
```

Its result is visualized in Fig. 3.13. It shows a circle somehow fitted in the set of the four co-linear points.

3.3 ANGLES AND DISTANCES

The inner product of geometric objects can be used in order to compute distances and angles between them (according to Table 2.4).

3.3.1 Distance Point-Line

The following listing

Listing 3.16 *DistancePointLine.clu*: Script for the computation of the distance between a point and a (normalized) line.

```
1  n1 = sqrt(2)/2;
2  n2 = sqrt(2)/2;
3  d = 1;
4
5  p1=2;
6  p2=1;
7
```

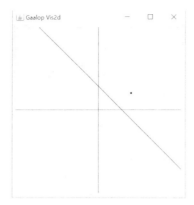

FIGURE 3.14 Visualization of *DistancePointLine.clu*: the computation of the distance between a point and a line.

```
8   P = createPoint(p1,p2);
9   L = n1*e1+n2*e2+d*einf;
10  ?Result = P.L;
11
12  :P;
13  :L;
```

visualizes a point and a (normalized) line acccording to Fig. 3.14. In line 10, the distance between the point and the line is computed as approximately 1.12 according to the following result shown in the output window

```
Result(0) = 1.121320343559643 // 1.0
```

Please refer to Sect. 7.2 for details.

3.3.2 Angle between two Lines

The inner product of lines can be used for the computation of the cosine of the angle between them according to Sect. 7.3. The angle of the two lines of Fig. 3.15, for instance, can be computed based on Listing 3.17.

Listing 3.17 *VisAngleBetweenLines.clu*: Script for the computation of the angle between two lines.

```
1   n1 = sqrt(2)/2;
2   n2 = sqrt(2)/2;
3   d = 1;
4
5   L1 = e1+d*einf;
6   L2 = n1*e1+n2*e2+d*einf;
```

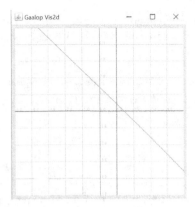

FIGURE 3.15 Visualization of *VisAngleBetweenLines.clu*: the computation of the angle between two (normalized) lines.

```
 7  ?Result = L1.L2;
 8  ?Angle = Acos(Result)*180/3.14159265359;
 9
10  :Red;
11  :L1;
12  :L2;
```

The result shown in the output window is

```
Result(0) = 0.7071067811865476 // 1.0
Angle(0) = 44.99999999999702 // 1.0
```

The angle between the two lines is 45 degrees, as expected.

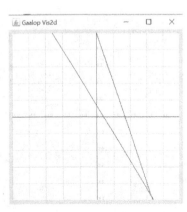

FIGURE 3.16 Visualization of *VisAngleBetweenLines2.clu*: the computation of the angle between two lines.

This computation is correct for normalized lines. But, what happens if the lines are not normalized as in the GAALOPScript according to Listing 3.18 (visualized in Fig. 3.16)?

Listing 3.18 *VisAngleBetweenLines2.clu*: Script for the computation of the angle between two (not normalized) lines.

```
1  p1  =  2;
2  p2  =  -1;
3  q1  =  1;
4  q2  =  2;
5
6  L1  =  *(createPoint(p1,p2)^createPoint(q1,q2)^einf);
7  L2  =  5*e1+3*e2+2*einf;
8  ?Result  =  L1.L2;
9
10 :Blue;
11 :L1;
12 :L2;
```

Looking at the output window

```
Result(0) = 18.0 // 1.0
```

we recognize, that the resulting value is out of the range of the *cos* function. The lines have to be normalized according to Listing 3.19 before applying the *acos* function.

Listing 3.19 *VisAngleBetweenLines3.clu*: Script for the computation of the angle between two lines.

```
1  p1  =  2;
2  p2  =  -1;
3  q1  =  1;
4  q2  =  2;
5
6  L1  =  *(createPoint(p1,p2)^createPoint(q1,q2)^einf);
7  L2  =  5*e1+3*e2+2*einf;
8  M1  =  L1/abs(L1);
9  M2  =  L2/abs(L2);
10 ?Result  =  M1.M2;
11 ?Angle  =  Acos(Result)*180/3.14159265359;
12
13 :Blue;
14 :M1;
15 :M2;
```

The normalization is done by scaling with the help of the *abs* function. Now, the result of the inner product is in the range of the cosine function

```
Result(0) = 0.9761870601839526 // 1.0
Angle(0) = 12.52880770915072 // 1.0
```

and the angle can be computed correctly.

3.3.3 Distance between two Circles

Listing 3.20 visualizes three circles according to Fig. 7.8.

Listing 3.20 *ThreeCircles.clu*: GAALOPScript for the visualization of three circles, C1 centered at the origin, C2 and C3 with center points along the x-axis.

```
1   r=1.5;
2
3   p1=0;
4   p2=0;
5   q1=2;
6   q2=0;
7
8   P = createPoint(p1,p2);
9   C1 = P - 0.5*r*r*einf;
10  Q = createPoint(q1,q2);
11  C2 = Q - 0.5*r*r*einf;
12  C3 = createPoint(-3.5,0) - 0.5*r*r*einf;
13  ?IPC1C2 = C1.C2;
14  ?IPC1C3 = C1.C3;
15
16  :Red;
17  :C1;
18  :Black;
19  :C2;
20  :C3;
```

In the first line, the variable for the radius r of the circles is defined. In the lines 3-6, the 2D coordinates of the center points of two of the circles are defined. The macro *createPoint(x,y)* computes in the lines 8 and 10 the 4D points P, Q based on these 2D coordinates. In the lines 9 and 11, the circles $C1, C2$ are computed. Line 12 computes the third circle $C3$ in one step. Lines 16-20 are responsible for visualization. This is indicated by leading colons. They describe colors and objects to be visualized accordingly.

In the lines 13 and 14, the inner products of the circle C1 with the circles C2 and C3 are computed, leading to the output

```
IPC1C2(0) = 0.25 // 1.0
IPC1C3(0) = -3.875 // 1.0
```

Please refer to Sect. 7.8 for the general treatment and to Eq. 7.24 for the formula for this example.

3.4 GEOMETRIC TRANSFORMATIONS

Table 3.3 summarizes how transformations according to Table 2.5 can be expressed in GAALOPScript.

TABLE 3.3 The GAALOPScript description of transformations of a geometric object o in Compass Ruler Algebra (note that e12 is the imaginary unit i).

	operator	Transformation
Reflection	L = n1*e1+n2*e2 + d* einf	-L*o*L
Rotation	R = cos (phi/2) - e12 * sin (phi/2)	R*o*(~R)
Translation	T = 1 - 0.5*(t1*e1+t2*e2)*einf	T*o*(~T)

3.4.1 Reflections

Reflections are the most basic operations in Compass Ruler Algebra. They can be easily expressed by the sandwich product

$$-LoL \tag{3.2}$$

of the reflection line L (see Sect. 3.2.3) and the geometric object o, which can be any object of the algebra, circle, line, point or point pair. Please notice that while for the geometric product no specific symbol is used, in GAALOP "*" is needed as a symbol. The following GAALOPScript describes the visualization of Fig. 3.17 with the geometric object o being a circle.

Listing 3.21 *ReflectCircle.clu*: Script for the visualization of the reflection of a circle.

```
1  x=1;
2  y=3;
3  r=1;
4
5  x1=0;
6  y1=-1;
7  x2=3;
8  y2=2;
9
10 :o = createPoint(x,y)-0.5*r*r*einf;
11 :L = *(createPoint(x1,y1)^createPoint(x2,y2)^einf);
```

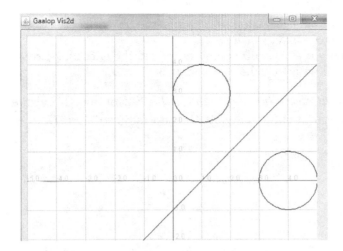

FIGURE 3.17 Visualization of *ReflectCircle.clu*: Reflection of a circle.

```
12  :oRefl = - L * o * L;
13  ?L;
14  ?o;
15  ?oRefl;
```

Computing the algebraic result leads to the reflected circle as

$$o_{\text{Refl}} = 72e_1 + 135e_\infty + 18e_0 \tag{3.3}$$

according to the following listing:

Listing 3.22 numerical output of *ReflectCircle.clu*: Script for the computation of the reflection of a circle.

```
1   L(1) = 3.0 // e1
2   L(2) = -3.0 // e2
3   L(3) = 3.0 // einf
4   o(1) = 1.0 // e1
5   o(2) = 3.0 // e2
6   o(3) = 4.5 // einf
7   o(4) = 1.0 // e0
8   oRefl(1) = 72.0 // e1
9   oRefl(3) = 135.0 // einf
10  oRefl(4) = 18.0 // e0
```

What we realize is, that the algebraic expression of the reflected circle is not normalized (its e_0-component is not 1). In order to normalize this circle, we have to divide it by 18 (see Sect. 5.6).

$$o_{\text{Refl}} = 4e_1 + 7.5e_\infty + e_0 \tag{3.4}$$

Now, the e_1- and e_2-components describe the correct 2D center point position of the reflected circle, which is (4,0).

If we use only normalized objects, we get the normalized result. This can be computed with the following Listing 3.23

Listing 3.23 *ReflectCircleNormalized.clu*: Script for the visualization of the reflection of the circle o at the (normalized) line L.

```
1  x=1;
2  y=3;
3  r=1;
4
5  x1=0;
6  y1=-1;
7  x2=3;
8  y2=2;
9
10 :o = createPoint(x,y)-0.5*r*r*einf;
11 L_notnormalized
12    = *(createPoint(x1,y1)^createPoint(x2,y2)^einf);
13 :L = L_notnormalized/abs(L_notnormalized);
14 :oRefl = - L * o * L;
15 ?L;
16 ?o;
17 ?oRefl;
```

Here, we first normalize the line L based on the *abs* function and then apply the reflection operation. The numerical result according to the following listing

Listing 3.24 numerical output of *ReflectCircleNormalized.clu*: Script for the computation of the reflection of a circle.

```
1  L(1) = 0.7071067811865476 // e1
2  L(2) = -0.7071067811865476 // e2
3  L(3) = 0.7071067811865476 // einf
4  o(1) = 1.0 // e1
5  o(2) = 3.0 // e2
6  o(3) = 4.5 // einf
7  o(4) = 1.0 // e0
8  oRefl(1) = 4.000000000000001 // e1
9  oRefl(3) = 7.500000000000002 // einf
10 oRefl(4) = 1.0 // e0
```

shows that both the line and the reflected circle are normalized.

As follows we will see how transformations such as rotations and translations can be expressed based on reflections.

3.4.1.1 Rotations based on reflections

Rotations can be expressed as two reflections with respect to two non-parallel lines. The following GAALOPScript describes the rotation of a circle. In Fig. 3.18 we realize that the rotation is a rotation around the intersection point of the two lines with an angle of twice the angle between the two lines.

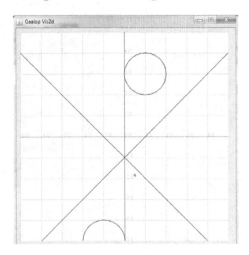

FIGURE 3.18 Visualization of *RotateCircle.clu*: Rotation of a circle around the intersection point of two lines.

Listing 3.25 *RotateCircle.clu*: Script for the visualization of the rotation of a circle based on two reflections.

```
1  x=1;
2  y=3;
3  r=1;
4
5  x1=0;
6  y1=-1;
7  x2=3;
8  y2=2;
9  x3=-2;
10 y3=1;
11
12 :o = createPoint(x,y)-0.5*r*r*einf;
13 :L = *(createPoint(x1,y1)^createPoint(x2,y2)^einf);
14 oRefl = - L * o * L;
15 :L2 = *(createPoint(x1,y1)^createPoint(x3,y3)^einf);
16 :oRefl2 = - L2 * oRefl * L2;
```

3.4.1.2   Translations based on reflections

Translations can be expressed as two reflections with respect to two parallel lines. Alternatively, they can be described based on translators according to Sect. 3.4.3.

3.4.1.3   Inversions

FIGURE 3.19   Visualization of *LineInversion.clu* (with t =2): the inversion of a line at a circle results in a circle.

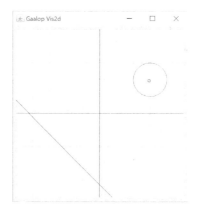

FIGURE 3.20   Visualization of *LineInversion.clu* (with t =-3): the inversion of a line at a circle results in a circle.

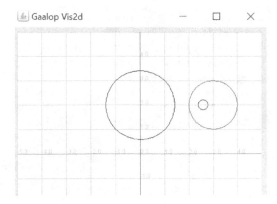

FIGURE 3.21 Visualization of *CircleInversion.clu* (with t =0): the inversion of a circle at a circle results in a circle.

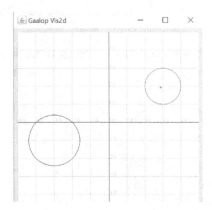

FIGURE 3.22 Visualization of *CircleInversion.clu* (with t =3): the inversion of a circle at a circle results in a circle.

Inversions are reflections not at lines, but at circles. Fig. 3.19 shows the inversion of a line at a (red) circle. Its result is another circle going through its center point.

This visualization is based on the following GAALOPScript:

Listing 3.26 *LineInversion.clu*: Inversion of a line at a circle.

```
1  x1 = 3;
2  y1 = 2;
3  r=1;
4  t=2;
5  x = sqrt(2)/2;
6  y = sqrt(2)/2;
```

```
 7
 8  L = x*e1+y*e2+t*einf;
 9  P1 = createPoint(x1,y1);
10  Ci = P1 - 0.5*r*r*einf;
11  Inversion = Ci*L*Ci;
12
13  :Green;
14  :P1;
15  :Red;
16  :Ci;
17  :Blue;
18  :L;
19  :Inversion;
```

If we change the parameter t (describing the distance of the line to the origin) from t=2 to t =-3, it results in another (smaller) circle also going through the origin, as shown in Fig. 3.20.

The following listing shows the inversion of a circle C at a circle Ci according to Fig. 3.21.

Listing 3.27 *CircleInversion.clu*: Inversion of a circle at another circle.

```
 1  x1 = 3;
 2  y1 = 2;
 3  t=0;
 4  r=1;
 5  x=-t;
 6  y=-t+2;
 7
 8  C = createPoint(x,y)-einf;
 9  P1 = createPoint(x1,y1);
10  Ci = P1 - 0.5*r*r*einf;
11  CircleInversion = Ci*C*Ci;
12
13  :Green;
14  :P1;
15  :Red;
16  :Ci;
17  :Blue;
18  :C;
19  :CircleInversion;
```

Fig. 3.22 shows the visualization if we change the parameter t from t=0 to t=3. It seems that with an increasing value of t the transformed circle gets more and more the center of C, which means it seems that the center of a circle can be computed based on the sandwich product

$$P = Ce_\infty C. \tag{3.5}$$

describing the inversion of infinity at the circle C. Please refer to Sect. 8.8 where we exactly show that.

3.4.2 Rotors

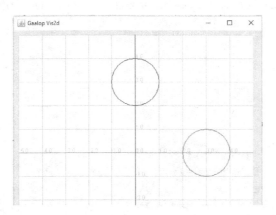

FIGURE 3.23 Visualization of *Rotor.clu*: the rotation of a circle (90 degrees) around the origin.

The following listing computes a rotor (for a rotation of 90 degrees), transforms a circle acccordingly and visualizes it in Fig. 3.23.

Listing 3.28 *Rotor.clu*: Script for the visualization of a rotation of a circle.

```
1   x=3;
2   y=0;
3   r=1;
4   angle=90;
5   alpha=(angle/180)*3.1416;
6   i = e1^e2;
7
8   P = createPoint(x,y);
9   Circle = P -0.5*r*r*einf;
10
11  Rota = cos(alpha/2) - i* sin(alpha/2);
12  Circle_rot = Rota * Circle * ~Rota;
13
14  :Red;
15  :Circle;
16  :Blue;
17  :Circle_rot;
```

By changing line 4, rotations of arbitrary angles can be realized. Please refer to Sect. 8.4 for the derivation of the rotor formula.

3.4.3 Translators

The following listing computes a translator, transforms a circle acccordingly and visualizes it in Fig. 3.24.

Listing 3.29 *Translator.clu*: Script for the visualization of a translation of a circle.

```
1  x=3;
2  y=0;
3
4  t1 = 1;
5  t2 = 1;
6  r=1;
7
8  P = createPoint(x,y);
9  Circle = P -0.5*r*r*einf;
10
11 T = 1-0.5*(t1*e1+t2*e2)^einf;
12 Circle_trans = T * Circle * ~T;
13
14 :Red;
15 :Circle;
16 :Blue;
17 :Circle_trans;
```

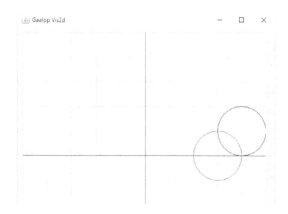

FIGURE 3.24 Visualization of *Translator.clu*: the translation of a circle with the translation vector (1,1).

Please refer to Sect. 8.5 for details of translators.

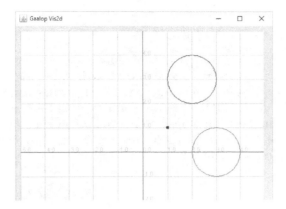

FIGURE 3.25 Visualization of *Motor.clu*: the rotation of a circle around the indicated point (90 degrees).

3.4.4 Motors

The following listing computes a combination of rotation and translation, transforms a circle acccordingly and visualizes it in Fig. 3.25.

Listing 3.30 *Motor.clu*: Script for the visualization of a combined rotation and translation of a circle.

```
1  x=3;
2  y=0;
3  r=1;
4  t1 = 1;
5  t2 = 1;
6  angle=90;
7  alpha=(angle/180)*3.1416;
8  i = e1^e2;
9
10 P = createPoint(x,y);
11 Circle = P -0.5*r*r*einf;
12
13 Rota = cos(alpha/2) - i* sin(alpha/2);
14 T = 1-0.5*(t1*e1+t2*e2)*einf;
15 Motor = T * Rota * ~T;
16 Circle_rot = Motor * Circle * ~Motor;
17
18 :Red;
19 :Circle;
20 :Blue;
21 :Circle_rot;
```

```
22 | :Black;
23 | :TP=createPoint(t1,t2);
```

Please refer to Sect. 8.6 for details of motors.

Congratulations, you are now able to work with Geometric Algebra as some kind of black box. If you are now interested in the mathematical background of what you did in SECTION I, please continue with the next SECTION II. If you are more interested in applications, you are able to directly switch to SECTION III with applications in the areas of robotics, computer vision and computer graphics.

II

Mathematical Foundations

CHAPTER 4

Mathematical Basics and 2D Euclidean Geometric Algebra

CONTENTS

Starting with this chapter the mathematical basics of Geometric Algebra are presented, here with the focus on 2D Euclidean Geometric Algebra, in the subsequent chapters with a focus on Compass Ruler Algebra.

4.1 THE BASIC ALGEBRAIC ELEMENTS OF GEOMETRIC ALGEBRA

While the pairwise orthogonal and normalized basis vectors $e_1, e_2, \ldots, e_n$ are the basic algebraic elements of an n-dimensional vector algebra, they are only one part of the algebraic elements of an n-dimensional Geometric Algebra[1]. **Blades** are the basic algebraic elements of Geometric Algebra. An n-dimensional Geometric Algebra consists of blades with **grades** 0, 1, 2, $\ldots$, n, where a scalar is a **0-blade** (a blade of grade 0) and the **1-blades** are the

---

[1]For simplicity, we use the term "n-dimensional Geometric Algebra" throughout this textbook. Although mathematically correct the term "2^n-dimensional geometric algebra of an n-dimensional vector space" should be used.

TABLE 4.1 The four basis blades of 2D Euclidean Geometric Algebra. This algebra consists of basic algebraic objects of grade (dimension) 0, the scalar, of grade 1 (the two basis vectors e_1 and e_2) and of grade 2 (the bivector $e_1 \wedge e_2$), which can be identified with the imaginary unit i squaring to -1.

Blade	Grade
1	0
e_1	1
e_2	1
$e_1 \wedge e_2$	2

basis vectors $e_1, e_2, \ldots, e_n$. The **2-blades**[2] $e_i \wedge e_j$ are blades spanned by two 1-blades, and so on. There exists only one element of the maximum grade n, $I = e_1 \wedge e_2 \ldots \wedge c_n$. It is therefore also called the **pseudoscalar**. A linear combination of k-blades is called a k-vector (or a vector, bivector, trivector. ...). A linear combination of blades with different grades is called a **multivector**. Multivectors are the general elements of a Geometric Algebra.

$G_{p,q,r}$ is an $n = p + q + r$-dimensional Geometric Algebra with the three different **signatures** 1, -1 and 0, which means with p basis vectors squaring to 1 ($e_i^2 = 1$), q basis vectors squaring to -1 and r basis vectors squaring to 0. There are 2^n blades in an n-dimensional Geometric Algebra.

An Euclidean Geometric Algebra G_n consists only of basis vectors squaring to 1. Table 4.1, for instance, shows the $4 = 2^2$ blades of the 2D Euclidean Geometric Algebra consisting of the scalar, two (basis) vectors and one bivector (the pseudoscalar of this algebra).

4.2 THE PRODUCTS OF GEOMETRIC ALGEBRA

The main product of Geometric Algebra is called the **geometric product**; many other products can be derived from it, especially the **outer** and the

TABLE 4.2 Notations for the Geometric Algebra products.

Notation	Meaning
AB	Geometric product of A and B
$A \wedge B$	Outer product of A and B
$A \cdot B$	Inner product of A and B

inner product. The notations of these products are listed in Table 4.2. Please

---

[2]Note that "∧" is the **outer product** as described in Section 4.2.

notice that no specific symbol is used in Geometric Algebra for the geometric product.

4.2.1 The Outer Product

Geometric Algebra provides an outer product ∧ with the properties listed in Table 4.3.

TABLE 4.3 Properties of the outer product ∧ of vectors.

Property	Meaning
Anti-Commutativity	$a \wedge b = -(b \wedge a)$
Distributivity	$a \wedge (b + c) = a \wedge b + a \wedge c$
Associativity	$a \wedge (b \wedge c) = (a \wedge b) \wedge c$

The outer product of two vectors a and b can be visualized as the parallelogram spanned by these two vectors according to Fig. 4.1. For a zero angle the outer product is zero. This is the reason why the outer product can be

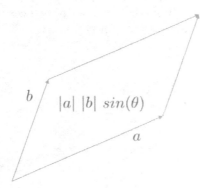

FIGURE 4.1 Magnitude of blade $a \wedge b$ is the area of the parallelogram spanned by a and b [57].

used as a measure of parallelness. Please refer to Chapt. 4 of [57] for more details.

In the case of 2D Euclidean Geometric Algebra

$$a \wedge b = |a| \, |b| \, sin(\theta) \, e_1 \wedge e_2, \tag{4.1}$$

means the outer product of two vectors a and b equals the area of the parallelogram spanned by a and b times the basis blade $e_1 \wedge e_2$. For normalized vectors n and m

$$n \wedge m = sin(\theta) e_1 \wedge e_2 \tag{4.2}$$

Computation example

We compute the outer product of two vectors:

$$c = (e_1 + e_2) \wedge (e_1 - e_2) \tag{4.3}$$

can be transformed based on distributivity to

$$c = (e_1 \wedge e_1) - (e_1 \wedge e_2) + (e_2 \wedge e_1) - (e_2 \wedge e_2); \tag{4.4}$$

since $u \wedge u = 0$,

$$c = -(e_1 \wedge e_2) + (e_2 \wedge e_1), \tag{4.5}$$

and because of anti-commutativity,

$$c = -(e_1 \wedge e_2) - (e_1 \wedge e_2) \tag{4.6}$$

or

$$c = -2(e_1 \wedge e_2). \tag{4.7}$$

4.2.2   The Inner Product

While the outer product is anti-commutative, the inner product is commutative. For Euclidean spaces, the inner product of two vectors is the same as the well-known Euclidean scalar product of two vectors.

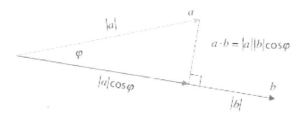

FIGURE 4.2   Scalar product of two vectors a and b.

As known from linear algebra, the following equation holds for arbitrary vectors a and b:

$$a \cdot b = |a| \, |b| \, cos(\theta) \tag{4.8}$$

and for normalized vectors n and m

$$n \cdot m = cos(\theta). \tag{4.9}$$

For perpendicular vectors, the inner product is 0, for instance,

$$e_1 \cdot e_2 = 0. \tag{4.10}$$

Please refer to Chapt. 3 of [57] for the general rule for the inner product of arbitrary multivectors. The inner product of a vector and a bivector, for instance, can be defined as

$$a \cdot (b \wedge c) = (a \cdot b)c - (a \cdot c)b. \tag{4.11}$$

The inner product of Geometric Algebra contains metric information. In Chapt. 7, we use Compass Ruler Algebra and its inner product for the computation of angles and distances.

4.2.3 The Geometric Product

The geometric product is an amazingly powerful operation, which is used mainly for the handling of transformations. The geometric product of vectors is a combination of the outer product and the inner product. The geometric product of u and v is denoted by uv (please notice that for the geometric product no specific symbol is used). For vectors u and v, the geometric product uv can be defined as the sum of outer and inner product

$$uv = u \wedge v + u \cdot v. \tag{4.12}$$

We derive the following for the inner and outer products:

$$u \cdot v = \frac{1}{2}(uv + vu), \tag{4.13}$$

$$u \wedge v = \frac{1}{2}(uv - vu), \tag{4.14}$$

but, as noted above, these formulas apply in this form only for vectors. See Sect. 3.1 of [57] for an axiomatic approach to the geometric product.
Computation example: What is the square of a vector?

$$u^2 = uu = \underbrace{u \wedge u}_{0} + u \cdot u = u \cdot u \tag{4.15}$$

for example

$$e_1^2 = e_1 \cdot e_1 = 1. \tag{4.16}$$

4.3 THE IMAGINARY UNIT IN GEOMETRIC ALGEBRA

Euclidean Geometric Algebra contains not only the two basis vectors e_1 and e_2 (as known from linear algebra), but also basis elements of grade (dimension) 0 and 2 (see Table 4.1). Grade 0 represents scalars and grade 2 represents

TABLE 4.4 Multiplication table of 2D Euclidean Geometric Algebra.

	1	e_1	e_2	$e_1 \wedge e_2$
1	1	e_1	e_2	$e_1 \wedge e_2$
e_1	e_1	1	$e_1 \wedge e_2$	e_2
e_2	e_2	$-e_1 \wedge e_2$	1	$-e_1$
$e_1 \wedge e_2$	$e_1 \wedge e_2$	$-e_2$	e_1	-1

the imaginary unit i. Its main property can be easily shown by the following calculation:

Since $e_1 e_2 = e_1 \wedge e_2 + \underbrace{e_1 \cdot e_2}_{0} = e_1 \wedge e_2,$

$$i^2 = (e_1 \wedge e_2)^2 = (e_1 e_2) \underbrace{(e_1 e_2)}_{-e_2 e_1} = -e_1 \underbrace{e_2 e_2}_{1} e_1 = -\underbrace{e_1 e_1}_{1} = -1 \qquad (4.17)$$

We realize that the element $e_1 \wedge e_2$ squares to -1. This is why linear combinations of the grade 0 and grade 2 elements of 2D Euclidean Geometric Algebra describe all complex numbers. Table 4.4 describes the multiplication table of all basis elements.

Complex numbers are one example of how Geometric Algebra subsumes other mathematical systems. The geometric meaning of these numbers are rotations in 2D. This is also true in 4D Compass Ruler Algebra as we will see in Sect. 6.2.

4.4 THE INVERSE

The **inverse** of a blade A is defined by

$$AA^{-1} = 1.$$

The inverse of a vector v, for instance, is

$$v^{-1} = \frac{v}{v \cdot v}.$$

Proof:

$$v \frac{v}{v \cdot v} = \frac{v \cdot v}{v \cdot v} = 1.$$

Example 1 The inverse of the vector $v = 2e_1$ results in $0.5e_1$, since $v \cdot v = 2$.

Example 2 The inverse of the (Euclidean) pseudoscalar $1/I$ is the negative of the pseudoscalar $(-I)$.

Proof:

$$II = (e_1 \wedge e_2)(e_1 \wedge e_2) = -1$$

$$\rightarrow II(I^{-1}) = -I^{-1}$$
$$\rightarrow I(II^{-1}) = -I^{-1}$$
$$\rightarrow I = -I^{-1}$$
$$\rightarrow I^{-1} = -I.$$

See [38] for details about multivector inverses.

4.5   THE DUAL

Since the geometric product is **invertible**, divisions by algebraic expressions are possible.

The **dual** of an algebraic expression is calculated by dividing it by the pseudoscalar I. In the following, the dual of the pseudoscalar $e_1 \wedge e_2$ is calculated. A superscript * means the dual operator.

$$(e_1 \wedge e_2)^* = (e_1 \wedge e_2)(e_1 \wedge e_2)^{-1}$$

$$(e_1 \wedge e_2)^* = (e_1 \wedge e_2)\underbrace{(e_1 \wedge e_2)^{-1}}_{-(e_1 \wedge e_2)}$$

$$(e_1 \wedge e_2)^* = -(e_1 \wedge e_2)(e_1 \wedge e_2)$$

$$(e_1 \wedge e_2)^* = -\underbrace{(e_1 \wedge e_2)(e_1 \wedge e_2)}_{-1}$$

$$(e_1 \wedge e_2)^* = 1.$$

See [57] for mathematical details.

4.6   THE REVERSE

The reverse of a multivector is the multivector with reversed order of the outer product components; for instance the reverse of $1 + e_1 \wedge e_2$ is $1 + e_2 \wedge e_1$ or $1 - e_1 \wedge e_2$.

CHAPTER 5

Compass Ruler Algebra and Its Geometric Objects

CONTENTS

The main advantage of Geometric Algebra is its easy and intuitive treatment of geometry. This is why the focus of this book is on the introduction of Geometric Algebra based on computing with the most basic geometric objects, namely points, lines and circles. While we are computing in 2D space, the underlying algebra is the 4D Compass Ruler Algebra which is simply the Conformal Geometric Algebra [17] [50] in 2D.

Chapt. 2 already presented Compass Ruler Algebra in a nutshell and we already worked with it in Chapt. 3. We realized, for instance, that circles can be defined based on the outer product of three points. This is why the circumcircle of a triangle can be expressed easily. In this chapter, we describe the algebraic structure as well as the geometric objects of Compass Ruler Algebra in some more detail. Here and in Chapt. 6, we will use the outer product for the construction and intersection of geometric objects, while the inner product will be used in Chapt. 7 for the computation of angles and distances and the geometric product in Chapt. 8 for the computation of transformations. For all our (symbolic) computations, we use the GAALOP software package. While, in SECTION I, we mainly used the visualization window, we are now mainly using the LaTeX code generator[1]. This code generator describes multivectors with its coefficients based on the indices of Table 2.2 as LaTeX code, that can be used directly in publications.

5.1 THE ALGEBRAIC STRUCTURE

The Compass Ruler Algebra $G_{3,1}$ uses the two Euclidean basis vectors e_1 and e_2 of the plane and two additional basis vectors e_+, e_- with positive and negative signatures, respectively, which means that they square to $+1$ as usual (e_+) and to -1 (e_-).

$$e_+^2 = 1, \qquad e_-^2 = -1, \qquad e_+ \cdot e_- = 0. \tag{5.1}$$

Another basis e_0, e_∞, with the following geometric meaning

e_0 represents the origin,

e_∞ represents infinity,

(see Sect. 5.9) can be defined with the relations

$$e_0 = \frac{1}{2}(e_- - e_+), \qquad e_\infty = e_- + e_+. \tag{5.2}$$

These new basis vectors are null vectors:

$$e_0^2 = e_\infty^2 = 0. \tag{5.3}$$

Taking their inner product results in

$$e_\infty \cdot e_0 = -1, \tag{5.4}$$

since

$$(e_- + e_+) \cdot \frac{1}{2}(e_- - e_+) = \frac{1}{2}(\underbrace{e_- \cdot e_-}_{-1} - \underbrace{e_- \cdot e_+}_{0} + \underbrace{e_+ \cdot e_-}_{0} - \underbrace{e_+ \cdot e_+}_{1}) = -1, \tag{5.5}$$

and their geometric product is

$$e_\infty e_0 = e_\infty \wedge e_0 + e_\infty \cdot e_0 = e_\infty \wedge e_0 - 1 \tag{5.6}$$

or

$$e_0 e_\infty = e_0 \wedge e_\infty + e_\infty \cdot e_0 = -e_\infty \wedge e_0 - 1. \tag{5.7}$$

Table 2.2 lists all the basis blades (outer products of subsets of the 4 basis

---

[1]See Sect. 3.1 for the configuration of GAALOP.

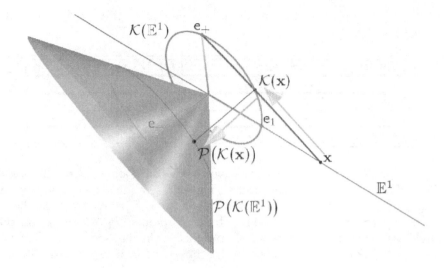

FIGURE 5.1 The mathematical model behind the Conformal Geometric Algebra of 1D space (image from [59]).

vectors). The algebraic basis elements of the algebra are the multivectors as a linear combination of these basis blades.

Compass Ruler Algebra is the Compass Ruler Algebra of the 2D space. Figure 5.1 shows the mathematical model behind the Conformal Geometric Algebra of 1D space. It illustrates the embedding of the 1-dimensional Euclidean space based on the basis vector e_1 in the 3D conformal space. First, the 1D space is embedded via a stereographic projection based on e_+ and second via a homogenization based on e_-. Please find details in the tutorial [59] and in the book [57].

5.2 THE BASIC GEOMETRIC ENTITIES AND THEIR NULL SPACES

In the Compass Ruler Algebra, geometric objects can be represented as algebraic expressions. Multivectors representing the basic geometric entities of the algebra, namely points, circles, lines, and point pairs, are listed in Table 5.1 (**x** and **n** are in bold type to indicate that they represent 2D entities obtained by linear combinations of the 2D basis vectors e_1 and e_2.).

TABLE 5.1 The representations of the geometric entities of the Compass Ruler Algebra.

Entity	IPNS representation	OPNS representation
Point	$P = \mathbf{x} + \frac{1}{2}\mathbf{x}^2 e_\infty + e_0$	
Circle	$C = P - \frac{1}{2}r^2 e_\infty$	$C^* = P_1 \wedge P_2 \wedge P_3$
Line	$L = \mathbf{n} + d e_\infty$	$L^* = P_1 \wedge P_2 \wedge e_\infty$
Point pair	$P_p = C_1 \wedge C_2$	$P_p^* = P_1 \wedge P_2$

L represents a line with normal vector $\mathbf{n}$ and distance d to the origin. The $\{Ci\}$ represent different circles. The outer product "$\wedge$" indicates the construction of a geometric object with the help of points $\{P_i\}$ that lie on it.[2] A circle, for instance, is defined by three points $(P_1 \wedge P_2 \wedge P_3)$ on this circle. Another meaning of the outer product is the intersection of geometric entities[3]. A point pair is defined by the intersection of two circles $C_1 \wedge C_2$. Since lines are specific circles with infinite radius (see Sect. 5.10), this is also true for the outer product of two lines as well as for the outer product of a circle and a line.

These entities have two algebraic representations: the (standard) **IPNS** (inner product null space) and the (dual) **OPNS** (outer product null space). The IPNS of the algebraic expression A are all the points X satisfying the equation

$$A \cdot X = 0. \tag{5.8}$$

The OPNS of the algebraic expression A are all the points X satisfying the equation

$$A \wedge X = 0. \tag{5.9}$$

These representations are duals of each other (a superscript asterisk denotes the dualization operator).

In the following, we present the representations of the basic geometric entities based on their null spaces.

5.3 POINTS

In order to represent points in Compass Ruler Algebra, the original 2D point

$$\mathbf{x} = x_1 e_1 + x_2 e_2 \tag{5.10}$$

is extended to a 4D vector by taking a linear combination of the 4D basis vectors e_1, e_2, e_∞, and e_0 according to the equation

$$P = \mathbf{x} + \frac{1}{2}\mathbf{x}^2 e_\infty + e_0. \tag{5.11}$$

---

[2]In the OPNS representation
[3]In the IPNS representation

where $\mathbf{x}^2$ is the inner product

$$\mathbf{x}^2 = (x_1 e_1 + x_2 e_2) \cdot (x_1 e_1 + x_2 e_2) = x_1^2 e_1^2 + 2 x_1 x_2 \underbrace{(e_1 \cdot e_2)}_{0} + x_2^2 e_2^2 = x_1^2 + x_2^2.$$

(5.12)

For example, for the 2D origin $(x_1, x_2) = (0,0)$ we get $P = e_0$.

In order to evaluate the geometric meaning of a point P with 2D coordinates (p_1, p_2), we compute its IPNS as its null space with respect to the inner product. The IPNS of P describes all the points X satisfying the equation

$$P \cdot X = 0.$$

(5.13)

The following GAALOPScript

Listing 5.1 *IPNSPoint.clu*: Computation of the IPNS of a point.

```
1  P = createPoint(p1,p2);
2  X = createPoint(x,y);
3  ?IPPoint =   P.X;
```

computes this inner product and assigns it to the variable *IPPoint* (GAALOP computes all the variables indicated by a leading question mark). This resulting multivector is equal to

$$IPPoint_0 = \frac{1}{2}(-y^2 + 2 * p_2 * y - x^2 + 2 * p_1 * x - p_2^2 - p_1^2)$$

(5.14)

with the null space

$$y^2 - 2 * p_2 * y + x^2 - 2 * p_1 * x + p_2^2 + p_1^2 = 0$$

or

$$(y - p_2)^2 + (x - p_1)^2 = 0$$

describing exactly the point P.

5.4 LINES

A line is defined by

$$L = \mathbf{n} + d e_\infty,$$

(5.15)

where $\mathbf{n} = n_1 e_1 + n_2 e_2$ refers to the 2D normal vector of the line L and d is the distance to the origin. The following GAALOPScript

Listing 5.2 *IPNSLine.clu*: Computation of the IPNS of a line.

```
1  X = createPoint(x,y);
2  L = n1*e1+n2*e2+d*einf;
3  ?IP = X.L;
```

computes the inner product of a line L and a general point X. This results in the IPNS

$$n_1 * x + n_2 * y - d = 0 \tag{5.16}$$

which is a line with the corresponding normal vector (n_1, n_2) and distance d to the origin.

A line can also be defined with the help of two points that lie on it and the point at infinity:

$$L^* = P_1 \wedge P_2 \wedge e_\infty. \tag{5.17}$$

Note that a line is a circle of infinite radius (see Sect 5.10).

5.5 CIRCLES

A circle can be represented with the help of its center point P and its radius r as

$$C = P - \frac{1}{2}r^2 e_\infty \tag{5.18}$$

or

$$C = \mathbf{x} + \frac{1}{2}\mathbf{x}^2 e_\infty + e_0 - \frac{1}{2}r^2 e_\infty \tag{5.19}$$

or

$$C = \mathbf{x} + \frac{1}{2}(\mathbf{x}^2 - r^2)e_\infty + e_0 \tag{5.20}$$

Note that the representation of a point is simply that of a circle of radius zero.

A circle can also be represented with the help of three points that lie on it, by

$$C^* = P_1 \wedge P_2 \wedge P_3. \tag{5.21}$$

As an example, we compute the IPNS of the expression $e_0 - \frac{1}{2}r^2 e_\infty$, which means all the points X satisfying the following equation

$$\left(e_0 - \frac{1}{2}r^2 e_\infty\right) \cdot X = 0. \tag{5.22}$$

The following GAALOPScript

Listing 5.3 *IPNSOriginCircle.clu*: Computation of the IPNS of an origin circle.

```
1  X = createPoint(x,y);
2  C = e0- 0.5*r*r*einf;
3  ?Result = C.X;
```

computes this inner product. The resulting multivector is equal to the scalar value $\frac{1}{2}(x^2 + y^2 - r^2)$ with the null space

$$x^2 + y^2 - r^2 = 0, \tag{5.23}$$

describing all points at the same distance r from the origin, namely a circle at the origin.

The following GAALOPScript

Listing 5.4 *CircleSquare.clu*: The square of a circle.

```
1  X = createPoint(x,y);
2  C = X - 0.5*r*r*einf;
3  ?CSquare = C*C;
```

computes the square of a circle and results in

$$CSquare_0 = r * r,$$

which means the square of a circle equals to the square of its radius or

$$r = \sqrt{C^2}. \tag{5.24}$$

5.6 NORMALIZED OBJECTS

Looking at the IPNS representations of point, circle and line of Table 5.1 we realize that they are all vectors of Compass Ruler Algebra. On the other hand, an arbitrary vector[4] must not have a representation as a geometric object. Considering an arbitrary vector

$$v = x_1 e_1 + x_2 e_2 + x_3 e_\infty + x_4 e_0 \tag{5.25}$$

and its null space

$$v \cdot X = 0 \tag{5.26}$$

we realize that

$$(cv) \cdot X = 0 \tag{5.27}$$

with an arbitrary scalar value $c \neq 0$ describing the same null space, since the IPNS equation $v \cdot X = 0$ is equivalent to the equation $(cv) \cdot X = c(v \cdot X) = 0$. This means that v and cv describe the same geometric object. Please notice that **this reasoning is not only true for vectors but also for arbitrary multivectors**, representing geometric objects.

With this knowledge, we are able to determine what the geometric meaning of an arbitrary vector v is. If its e_0-component is zero, it represents a line

$$L = x_1 e_1 + x_2 e_2 + x_3 e_\infty. \tag{5.28}$$

---

[4]Arbitrary linear combinations of the basis vectors e_1, e_2, e_0, e_∞

Its normalized form can be computed by scaling with the length of the 2D vector (x_1, x_2), which can be expressed as

$$L_{normalized} = \frac{L}{|L|}. \tag{5.29}$$

This can be shown based on the following GAALOPScript

Listing 5.5 *normalizeLine.clu*: Normalization of a line.

```
1  line = n1*e1+n2*e2+n3*einf;
2  L = k*line;
3  ?LAbs = abs(L);
```

resulting in

$$LAbs_0 = \sqrt{k * k * n2 * n2 + k * k * n1 * n1}$$

or

$$LAbs_0 = \sqrt{k * k * \underbrace{(n2 * n2 + n1 * n1)}_{1}}$$

or

$$LAbs_0 = k.$$

In the case of points and circles, the e_0-component equals to 1. This is why an arbitrary vector v has to be scaled by $x_4 \neq 0$.

$$C = \frac{x_1}{x_4}e_1 + \frac{x_2}{x_4}e_2 + \frac{x_3}{x_4}e_\infty + e_0. \tag{5.30}$$

This can be done based on the formula

$$C = -\frac{v}{v.e_\infty} \tag{5.31}$$

which can be shown based on the following GAALOPScript

Listing 5.6 *normalizeCircle.clu*: Normalization of a circle.

```
1  v = x1*e1+x2*e2+x3*einf+x4*e0;
2  ?C = -v/(v.einf);
```

with the result

$$vnormalized_1 = \frac{x1}{x4}$$

$$vnormalized_2 = \frac{x2}{x4}$$

$$vnormalized_3 = \frac{x3}{x4}$$

$$vnormalized_4 = 1$$

5.7 THE DIFFERENCE OF TWO POINTS

In our example of Sect. 3.2.6 the difference of two points computes the line in the middle of two points. Is that true in arbitrary cases? We will show that in Sect. 16.3.

5.8 THE SUM OF POINTS

In Sect. 3.2.7, we realized that the sum of points can be used for some kind of fitting a circle into a set of points. In the case of four co-linear points, for instance, the result is visualized in Fig. 3.13. It also shows a circle somehow fitted in the set of the four points. But, in many applications we are interested in a result better accommodating that the points are lying on one straight line. Please refer to Chapt. 13 for the approach of computing the best-fitting line or circle into a set of points.

5.9 THE MEANING OF E_0 AND E_∞

In order to evaluate the geometric meaning of e_0, we are able to compute its IPNS as its null space with respect to the inner product. The IPNS of e_0 describes all the points X satisfying the equation

$$e_0 \cdot X = 0. \tag{5.32}$$

The following GAALOPScript

Listing 5.7 *IPNSe0.clu*: Computation of the IPNS of e_0.

```
1  X = createPoint(x,y);
2  ?Result = e0.X;
```

computes this inner product and assigns it to the variable *Result* (GAALOP computes all the variables indicated by a leading question mark). This resulting multivector is equal to the scalar value $x^2 + y^2$ with the null space

$$x^2 + y^2 = 0, \tag{5.33}$$

describing exactly the point at the origin.

In order to evaluate the geometric meaning of e_∞, we assume an arbitrary Euclidean point $\mathbf{x} = x_1 e_1 + x_2 e_2$ (not equal to the origin) with a normalized Euclidean vector $\mathbf{n}$ in the direction of $\mathbf{x}$,

$$\mathbf{x} = t\mathbf{n}, \ t > 0, \ \mathbf{n}^2 = 1 \tag{5.34}$$

with its representation P according to equation (5.11) and consider its limit $\lim_{t \to \infty}$. Another (homogeneous) representation of this point P is cP, its product with an arbitrary scalar value $c \neq 0$ (see Sect. 5.6).

Let us choose the arbitrary scalar value as $c = \frac{2}{\mathbf{x}^2}$ and consider $P' = \frac{2}{\mathbf{x}^2} P$,

$$P' = \frac{2}{\mathbf{x}^2} (\mathbf{x} + \frac{1}{2} \mathbf{x}^2 e_\infty + e_0), \tag{5.35}$$

$$P' = \frac{2}{\mathbf{x}^2} \mathbf{x} + e_\infty + \frac{2}{\mathbf{x}^2} e_0. \tag{5.36}$$

We use this form to compute the limit $\lim_{t \to \infty} P'$ for increasing $\mathbf{x}$. Since $\mathbf{x} = t\mathbf{n}$, we get

$$P' = \frac{2}{t^2 \mathbf{n}^2} t\mathbf{n} + e_\infty + \frac{2}{t^2 \mathbf{n}^2} e_0 \tag{5.37}$$

and, since $\mathbf{n}^2 = 1$,

$$P' = \frac{2}{t} \mathbf{n} + e_\infty + \frac{2}{t^2} e_0. \tag{5.38}$$

Based on this formula and the fact that P and P' represent the same Euclidean point, we can easily see that the point at infinity for any direction vector $\mathbf{n}$ is represented by e_∞:

$$\lim_{t \to \infty} P' = e_\infty. \tag{5.39}$$

5.10 LINE AS A LIMIT OF A CIRCLE

Circles and lines are both vectors in Compass Ruler Algebra. In this section, we will see how a circle

$$C = \mathbf{c} + \frac{1}{2} (\mathbf{c}^2 - r^2) e_\infty + e_0, \tag{5.40}$$

with an Euclidean center point $\mathbf{c}$ and radius r, degenerates to a line as the result of a limiting process.

According to the construction shown in Fig. 5.2, the minimum distance from the origin to a line with its center in the direction opposite to a normal vector $\mathbf{n}$ is

$$d = r - \sqrt{\mathbf{c}^2}, \tag{5.41}$$

and the radius is the sum of the length of the 2D vector $\mathbf{c}$ and d, i.e.,

$$r = \sqrt{\mathbf{c}^2} + d, \tag{5.42}$$

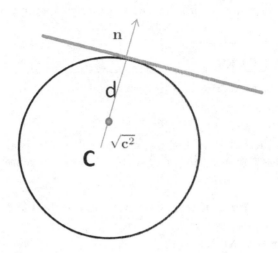

FIGURE 5.2 A circle with a center **c** (in the direction opposite to a normal vector **n**) with radius that goes to infinity (while the radius of the circle changes accordingly), results finally in a line with the normal vector **n** and a distance d from the origin.

or

$$r^2 = \mathbf{c}^2 + 2d\sqrt{\mathbf{c}^2} + d^2. \tag{5.43}$$

The circle can be written as

$$C = \mathbf{c} + \frac{1}{2}(\mathbf{c}^2 - \mathbf{c}^2 - 2d\sqrt{\mathbf{c}^2} - d^2)e_\infty + e_0 \tag{5.44}$$

or, equivalently,

$$C = \mathbf{c} + \frac{1}{2}(-2d\sqrt{\mathbf{c}^2} - d^2)e_\infty + e_0. \tag{5.45}$$

Now we introduce C', a scaled version of the algebraic expression for the circle C representing geometrically the same circle, as follows:

$$C' = -\frac{C}{\sqrt{\mathbf{c}^2}} = -\frac{\mathbf{c}}{\sqrt{\mathbf{c}^2}} + \frac{1}{2}\left(2d + \frac{d^2}{\sqrt{\mathbf{c}^2}}\right)e_\infty - \frac{e_0}{\sqrt{\mathbf{c}^2}}. \tag{5.46}$$

Since the ratio of the 2D vector **c** to its length $\sqrt{\mathbf{c}^2}$ corresponds to the negative normal vector **n** (see the construction in Fig. 5.2),

$$\lim_{\mathbf{c}^2 \to \infty}\left(-\frac{C}{\sqrt{\mathbf{c}^2}}\right) = \mathbf{n} + \lim_{\mathbf{c}^2 \to \infty}\frac{1}{2}\left(2d + \frac{d^2}{\sqrt{\mathbf{c}^2}}\right)e_\infty - \lim_{\mathbf{c}^2 \to \infty}\frac{e_0}{\sqrt{\mathbf{c}^2}}. \tag{5.47}$$

This is equivalent to

$$\lim_{\mathbf{c}^2 \to \infty}\left(-\frac{C}{\sqrt{\mathbf{c}^2}}\right) = \mathbf{n} + de_\infty, \tag{5.48}$$

which is a representation of a line with a normal vector $\mathbf{n}$ and a distance d from the origin.

5.11 POINT PAIRS

Point pairs can be represented directly by the dual of the outer product of two points

$$Pp^* = P_1 \wedge P_2. \tag{5.49}$$

Based on the following Listing 5.8, we will verify that the IPNS of an algebraic expression according to Eq. (5.49) is really representing a pair of points.

Listing 5.8 *IPNSPointPair.clu*: Computation of the IPNS of a point pair.

```
1  P  =  createPoint (x,y);
2  P1  =  createPoint (x1,y1);
3  P2  =  createPoint (x2,y2);
4  OuterP1P2  =  2*P1^P2;
5  PP  =  *OuterP1P2;
6  ?IP_PP  =   PP.P;
```

The resulting multivector of the inner product of an arbitrary point P with the dual of the outer product of two points $P1$ and $P2$ is described in Listing 5.9.

Listing 5.9 *Result IPNSPointPair.clu*: the computation of the IPNS of a point pair.

```
1   IP_PP [1] = ((y-y1)*y2*y2+((y1*y1-y*y+x1*x1)-x*x) *y2)
2    -y*y1*y1 + (y*y-x2*x2+x*x) *y1 +(x2*x2-x1*x1)* y;//e1
3   IP_PP [2] = ((x1-x)*y2*y2 (x-x2)*y1*y1 + (x2- x1) *y*y
4    +(x1-x)*x2*x2+(x*x-x1*x1)*x2+x*x1*x1) -x*x*x1; //e2
5   IP_PP [3] = ((x1*y-x*y1)*y2*y2
6    +((x*y1*y1-x1*y*y+x*x1*x1)-x*x*x1) *y2)-x2*y*y1*y1
7    +(x2*y*y-x*x2*x2+x*x*x2)*y1
8    +(x1*x2*x2-x1*x1*x2)*y;//einf
9   IP_PP [4] = (2.0*x1-2.0*x)*y2+(2.0*x-2.0*x2)*y1
10   +(2.0*x2-2.0*x1)*y; // e0
```

For the IPNS of this multivector we have to compute the set of points where all of its four coefficients are zero. Solving this algebraic equation system with Maxima leads to $[[x = x2, y = y2], [x = x1, y = y1]]$, which means a set of two points represented by the points $P1$ and $P2$.

A point pair can also be defined by the intersection of two circles (one or both circles can also be lines, which are specific circles according to Sect. 5.10)

$$Pp = C_1 \wedge C_2 \tag{5.50}$$

We can use the following formula to extract the two points of the point pair Pp (see [8, 12]):

$$P_{1,2} = \frac{Pp^* \pm \sqrt{Pp^* \cdot Pp^*}}{e_\infty \cdot Pp^*} \qquad (5.51)$$

or

$$P_1 = \frac{Pp^* + \sqrt{Pp^* \cdot Pp^*}}{e_\infty \cdot Pp^*} \qquad (5.52)$$

and

$$P_2 = \frac{Pp^* - \sqrt{Pp^* \cdot Pp^*}}{e_\infty \cdot Pp^*}. \qquad (5.53)$$

In case the points do not have to be normalized, the following multiplication

$$P_{1,2} = (Pp^* \pm \sqrt{Pp^* \cdot Pp^*})(e_\infty \cdot Pp^*) \qquad (5.54)$$

can be used instead of the division.

Please refer to Chapter 6 for more details about point pairs.

Intersections in Compass Ruler Algebra

CONTENTS

Intersections of geometric objects can be easily expressed in Geometric Algebra based on the outer product[1]. In Compass Ruler Algebra, intersections result in some kind of point pairs. In Sect. 3.2.4, for instance, the intersection of two circles is computed and visualized. Please notice, that in 3D, circles, lines and point pairs can result from intersection operations (see Chapter 15).

6.1   THE IPNS OF THE OUTER PRODUCT OF TWO VECTORS

First of all, we will show that the outer product of two vectors A and B really represents the intersection of the two objects represented by A an B. The IPNS of $A \wedge B$ is defined as the set of points X satisfying the following equation (see Sect. 5.6)

$$X \cdot (A \wedge B) = 0 \qquad (6.1)$$

which is equivalent to

$$(X \cdot A)B - (X \cdot B)A = 0 \qquad (6.2)$$

according to the rules of the inner product of a vector and a bivector (see Sect. 4.2.2). This can only be zero if

$$X \cdot A = 0 = X \cdot B \qquad (6.3)$$

---

[1]In the standard IPNS representation

which means the IPNS of $A \wedge B$ equals all the points belonging to A and B.

6.2 THE ROLE OF $E_1 \wedge E_2$

In Sect. 4.3, we realized that for $i = e_1 \wedge e_2$

$$i^2 = (e_1 \wedge e_2)^2 = -1. \tag{6.4}$$

This computation is also true for Compass Ruler Algebra. While i algebraically can be used as the imaginary unit, the question arises regarding what its geometric meaning is.

TABLE 6.1 The two representations of point pairs.

IPNS representation	OPNS representation
$P_p = C_1 \wedge C_2$	$P_p^* = P_1 \wedge P_2$

According to Table 6.1, point pairs have two representations based on the outer product: the OPNS representation is based on the outer product of two points, while the IPNS representation is based on the intersection of lines/circles generating point pairs. The two representations are duals to each other.

What is the role of $e_1 \wedge e_2$ then? Since e_1 and e_2 represent two lines through the origin with normals e_1 and e_2, $e_1 \wedge e_2$ represents the intersection of these lines. Taking its dual

```
?PP = *(e1^e2);
```

with GAALOP results in

```
PP[10] = 1.0; // einf ^ e0
```

which means

$$(e_1 \wedge e_2)^* = e_\infty \wedge e_0, \tag{6.5}$$

which is the outer product of two specific points, namely the origin and infinity.

Please refer to Sect. 8.2 for the role of $e_1 \wedge e_2$ with regard to transformations.

6.3 THE INTERSECTION OF TWO LINES

In the previous section, we realized that the intersection of the two specific lines e_1 and e_2 through the origin corresponds to a point pair of origin (their intersection point) and infinity. Here, we will see that this is true also for arbitrary lines, meaning that the result of the intersection of two arbitrary lines is a point pair of the real intersection point and the point of infinity.

The following GAALOPScript

Listing 6.1 *IntersectLines.clu*: Computation of the intersection of lines.

```
1  P = createPoint(p1,p2);
2  PPdual = P^einf;
3  ?PP = *PPdual;
4
5  L = l1*e1+l2*e2+l3*einf;
6  M = m1*e1+m2*e2+m3*einf;
7
8  ?I = L^M;
```

at first computes this kind of point pair as the dual of the outer product of a point and e_∞ resulting in

$$PP = (P \wedge e_\infty)^* = e_1 \wedge e_2 + p_2 e_1 \wedge e_\infty - p_1 e_2 \wedge e_\infty \qquad (6.6)$$

and then the intersection of the lines L and M resulting in

$$I = (l_1 m_2 - l_2 m_1)e_1 \wedge e_2 + (l_1 m_3 - l_3 m_1)e_1 \wedge e_\infty + (l_2 m_3 - l_3 m_2)e_2 \wedge e_\infty \quad (6.7)$$

or in scaled form

$$I_{scaled} = e_1 \wedge e_2 + \frac{l_1 m_3 - l_3 m_1}{l_1 m_2 - l_2 m_1} e_1 \wedge e_\infty + \frac{l_2 m_3 - l_3 m_2}{l_1 m_2 - l_2 m_1} e_2 \wedge e_\infty \qquad (6.8)$$

Comparing the coefficients of the multivectors PP and I_{scaled}, we recognize that the real intersection point of two lines L and M can be computed as follows:

$$p_1 = -\frac{l_2 m_3 - l_3 m_2}{l_1 m_2 - l_2 m_1} \qquad (6.9)$$

$$= \frac{l_3 m_2 - l_2 m_3}{l_1 m_2 - l_2 m_1} \qquad (6.10)$$

$$p_2 = \frac{l_1 m_3 - l_3 m_1}{l_1 m_2 - l_2 m_1} \qquad (6.11)$$

This means that the intersection of two lines is represented by a point pair of the point with the 2D coordinates (p_1, p_2) and the point at infinity.

6.4   THE INTERSECTION OF TWO PARALLEL LINES

The following GAALOPScript

Listing 6.2 *IntersectParallelLines.clu*: Computation of the intersection of parallel lines.

```
1  n = n1*e1 + n2*e2;
2  L1 = n + d1*einf;
3  L2 = n + d2*einf;
4  ?IL = L1^L2;
```

computes the intersection of two lines with the same normal vector but different distances to the origin. It results in

$$\mathbf{v} \wedge e_\infty \qquad (6.12)$$

(this is called a free vector in the literature [8]) with

$$\mathbf{v} = (d_2 - d_1)\mathbf{n}. \qquad (6.13)$$

$(d_2 - d_1)$ describes the distance between the two lines with normal vector $\mathbf{n}$. This is why $\mathbf{v}$ describes the translation vector in order to translate one line into the other.

6.5 THE INTERSECTION OF CIRCLE-LINE

We will use the intersection of a circle and a line in order to analyze the geometric meaning of an arbitrary point pair. The following GAALOPScript

Listing 6.3 *PointPairFromCircleandLine.clu*: computation of a point pair.

```
1  C  =  createPoint(c1,c2)-0.5*r*r*einf;
2  c  =  c1*e1+c2*e2;
3  n  =  n1*e1+n2*e2;
4  d  =  n.c;
5  L  =  n + d*einf;
6  ?PP  =  2*(C^L);
```

computes a point pair based on the intersection of a circle C with a line L going through the center point of the circle.

It results in the following C code

Listing 6.4 Result of the computation of a point pair according to *Point-PairFromCircleandLine.clu*.

```
1  void calculate(float c1, float c2, float n1, float n2,
2                 float r, float PP[16]) {
3
4  PP[5]  =  2.0 * c1 * n2 - 2.0 * c2 * n1; // e1^e2
5  PP[6]  =  n1 * r * r + 2.0 * c1 * c2 * n2
6             + (c1 * c1 - c2 * c2) * n1; // e1^einf
7  PP[7]  =  (-(2.0 * n1)); // e1 ^ e0
8  PP[8]  =  n2 * r * r + (c2 * c2 - c1 * c1) * n2
9             + 2.0 * c1 * c2 * n1; // e2^einf
10 PP[9]  =  (-(2.0 * n2)); // e2 ^ e0
11 PP[10] =  (-(2.0 * c2 * n2)) - 2.0 * c1 * n1; //einf^e0
12 }
```

This can be expressed in the following formula[2]:

$$-\frac{1}{2}Pp = -|\mathbf{c} \ \mathbf{n}|e_{12} + \mathbf{n} \wedge e_0 + (\mathbf{c} \cdot \mathbf{n})(e_\infty \wedge e_0) \quad (6.14)$$

$$+ \left[\frac{1}{2}(\mathbf{c}^2 - r^2)\mathbf{n} - (\mathbf{c} \cdot \mathbf{n})\mathbf{c}\right] \wedge e_\infty.$$

This representation brings about the possibility to extract the normal $\mathbf{n}$ and the center point $\mathbf{c}$ (as well as the radius r) of the point pair multivector. The normal vector $\mathbf{n}$ is a standard Euclidian vector with blades e_1 and e_2. The operation $\mathbf{n} \wedge e_0$ integrates the normal as linear combination $n_1 * e_1 \wedge e_0 + n_2 * e_2 \wedge e_0$ into the point pair multivector. It is therefore trivial to retrieve the normal, by accessing the coefficients of blades $e_1 \wedge e_0$, and $e_2 \wedge e_0$ (based on the normal vector we are able to normalize the point pair by dividing by the length of this vector). Retrieving the center point $\mathbf{c}$ as well as the radius r is slightly more complex. One possible solution is to consider Listing 6.4 as an equation system with the known point pair coefficients and let Maxima compute the unknown $c1, c2, n1, n2$ and r.

6.6 ORIENTED POINTS

What happens if the radius of the circle of Listing 6.3 equals zero? According to Eq. (6.14) the result is the following multivector

$$O_p = -|\mathbf{c} \ \mathbf{n}|e_{12} + \mathbf{n} \wedge e_0 + (\mathbf{c} \cdot \mathbf{n})(e_\infty \wedge e_0) + \left[\frac{1}{2}\mathbf{c}^2\mathbf{n} - (\mathbf{c} \cdot \mathbf{n})\mathbf{c}\right] \wedge e_\infty. \quad (6.15)$$

We call this kind of geometric object an *oriented point*[3]. It is a very interesting object, since it represents a point consisting of an orientation defined by the normal vector $\mathbf{n}$.

6.7 THE INTERSECTION OF CIRCLES

One may define a point pair as the intersection of two arbitrary circles with 2D center points $\mathbf{c_i} = c_{ix}e_1 + c_{iy}e_2$ and radii r_i. This is expressed in the following GAALOPScript

Listing 6.5 *computeCircleCircleCut.clu*: Computation of the intersection of two circles.

```
1  C1 = createPoint(c1x,c1y)-0.5*r1*r1*einf;
2  C2 = createPoint(c2x,c2y)-0.5*r2*r2*einf;
3  ?PP=2*C1^C2;
```

---

[2]$|\mathbf{c} \ \mathbf{n}|$ means the determinant of the 2D vectors $\mathbf{c}$ and $\mathbf{n}$.
[3]Please find details in [28].

resulting in

Listing 6.6 computeCircleCircleCut.c: Computation of the intersection of two circles.

```
 1  P[5] = 2.0 * c1x * c2y - 2.0 * c1y * c2x; // e1 ^ e2
 2  PP[6] = (-c1x) * r2 * r2 + c2x * r1 * r1
 3        + c1x * c2y * c2y + c1x * c2x * c2x
 4        + (-(c1y * c1y) - c1x * c1x) * c2x; // e1 ^ einf
 5  PP[7] = 2.0 * c1x - 2.0 * c2x; // e1 ^ e0
 6  PP[8] = (-c1y) * r2 * r2 + c2y * r1 * r1
 7        + c1y * c2y * c2y -c1y * c1y - c1x * c1x * c2y
 8        + c1y * c2x * c2x; // e2^einf
 9  PP[9] = 2.0 * c1y - 2.0 * c2y; // e2 ^ e0
10  PP[10] = r2 * r2 - r1 * r1 - c2y * c2y
11         - c2x * c2x + c1y * c1y + c1x * c1x; // einf^e0
```

or

$$Pp = -|\mathbf{c_1}\ \mathbf{c_2}|e_{12} + (\mathbf{c_1} - \mathbf{c_2}) \wedge e_0 \tag{6.16}$$

$$+\frac{1}{2}\left[(\mathbf{c_1}^2 - r_1^2) - (\mathbf{c_2}^2 - r_2^2)\right](e_\infty \wedge e_0)$$

$$-\frac{1}{2}\left[(\mathbf{c_1}^2 - r_1^2)\mathbf{c_2} - (\mathbf{c_2}^2 - r_2^2)\mathbf{c_1}\right] \wedge e_\infty.$$

CHAPTER 7

Distances and Angles in Compass Ruler Algebra

CONTENTS

In this section, we use Compass Ruler Algebra and its inner product for the computation of angles and distances. Original work dealing with inner products of circles derives from Jakob Steiner in the 1820's [A1][1]. There is more recent accessible content where the formulas and meaning for the inner products of circles appear in the "Kreisgeometrie" ("Circle geometry") section of Felix Klein's classic work "Vorlesungen ueber Hoehere Geometrie" [A2][2]. Other work (e.g. [A3][3].) also includes a comprehensive treatment with topics such as the angle of intersection of two intersecting circles and relations between circles and lines.

The goal of this section is to show how easy it is to derive this kind of result based on the GAALOP tool (see Sect. 3.1). We will in detail investigate

---

[1][A1]Jakob Steiner. Einige geometrische Betrachtungen . Journal fuer die reine und angewandte Mathematik, 1826, 1: 1-62

[2][A2] Felix Klein. Kreisgeometrie. Vorlesungen ueber Hoehere Geometrie. Springer, 1926:39

[3][A3] M. Berger. Geometry II, Springer, 1987

the inner product between lines, circles and points. The inner product of this kind of object is a scalar and can be used as a measure of distance (or as a description of the angle between two lines) as summarized in Table 7.1.

TABLE 7.1 Geometric meaning of the inner product of (normalized) lines, circles and points.

·	Line	Circle	Point
Line	Angle between lines Eq. (7.9)	Euclidean distance from center, Eq. (7.13)	Euclidean distance Eq. (7.6)
Circle	Euclidean distance from center, Eq. (7.13)	Distance measure Fig. 7.7	Distance measure Eq. (7.16)
Point	Euclidean distance Eq. (7.6)	Distance measure Eq. (7.16)	Distance Eq. (7.3)

In the following examples, we will see that the inner product $P \cdot Q$ of two geometric objects (represented by the vectors P and Q) can be used for tasks such as

- the distance between two points,

- the distance between a point and a line,

- the decision as to whether a point is inside or outside a circle,

etc.

7.1 DISTANCE BETWEEN POINTS

The following GAALOPScript computes the inner product of two points P and Q

Listing 7.1 *DistancePointPoint.clu*: Computation of the inner product of two points.

```
1  P = createPoint(p1,p2);
2  Q = createPoint(q1,q2);
3  ?Result = P.Q;
```

and results in

$$Result_0 = -0.5 * q2 * q2 + p2 * q2 - \frac{q1 * q1}{2} + p1 * q1 - \frac{p2 * p2}{2} - \frac{p1 * p1}{2}$$

or

$$P \cdot Q = -\frac{1}{2}(q_2^2 - 2p_2q_2 + q_1^2 + p_2^2 - 2p_1q_1 + p_1^2) \tag{7.1}$$

$$= -\frac{1}{2}((q_1 - p_1)^2 + (q_2 - p_2)^2) \tag{7.2}$$

$$= -\frac{1}{2}(\mathbf{q} - \mathbf{p})^2 \tag{7.3}$$

We recognize that the square of the Euclidean distance of the 2D points corresponds to the inner product of the 4D representation of the points multiplied by -2.

$$(\mathbf{q} - \mathbf{p})^2 = -2(P \cdot Q) \tag{7.4}$$

7.2 DISTANCE BETWEEN A POINT AND A LINE

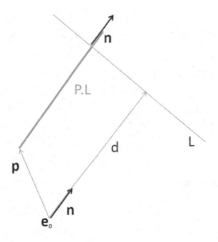

FIGURE 7.1 The inner product of a point and a line describes the distance between them.

The following GAALOPScript computes the inner product of the point P and the (normalized) line L (see Sect. 5.6 for normalized objects).

Listing 7.2 *DistancePointLine.clu*: Computation of the inner product of a point and a line.

```
1  P = createPoint(p1,p2);
2  L = n1*e1+n2*e2+d*einf;
3  ?Result = P.L;
```

and results in

$$P \cdot L = p_1 n_1 + p_2 n_2 - d \tag{7.5}$$

$$= \mathbf{p} \cdot \mathbf{n} - d, \tag{7.6}$$

which represents the Euclidean distance between the point and the line, with a sign according to

$P \cdot L > 0$: $\mathbf{p}$ is on the normal $\mathbf{n}$ side of the line;

$P \cdot L = 0$: $\mathbf{p}$ is on the line;

$P \cdot L < 0$: $\mathbf{p}$ is on the opposite side of the normal $\mathbf{n}$.

7.3 ANGLES BETWEEN LINES

Let us now derive an expression for the angle between two lines. The inner product of the line $L_1 = \mathbf{n_1} + d_1 e_\infty$ with normal vector $\mathbf{n_1}$ and distance d_1 and another line $L_2 = \mathbf{n_2} + d_2 e_\infty$ can be computed with the help of the following GAALOPScript

Listing 7.3 *AngleBetweenNormalizedLines.clu*: Computation of the inner product of two lines.

```
1  L1 = n11*e1+n12*e2+d1*einf;
2  L2 = n21*e1+n22*e2+d2*einf;
3  ?Result = L1.L2;
4  ?ResultDualLines = *L1.*L2;
```

resulting in

$$L_1 \cdot L_2 = \mathbf{n_1} \cdot \mathbf{n_2} \tag{7.7}$$

as well as for the dual lines

$$L_1^* \cdot L_2^* = \mathbf{n_1} \cdot \mathbf{n_2}. \tag{7.8}$$

Both, the inner product of two lines and the inner product of their duals describe the scalar product of the two normals of the lines. Based on this observation, the angle θ between the two lines can be computed as

$$\cos(\theta) = L_1 \cdot L_2 \tag{7.9}$$

or

$$\cos(\theta) = L_1^* \cdot L_2^*. \tag{7.10}$$

This is true for normalized lines. For not-normalized lines as defined in the following GAALOPScript

Listing 7.4 *AngleBetweenLines.clu*: Computation of the inner product of two (not normalized) lines.

```
1  L1 = l1*(n1x*e1+n1y*e2+d1*einf);
2  L2 = l2*(n2x*e1+n2y*e2+d2*einf);
3  ?Result = L1.L2;
```

the inner product of the lines is multiplied by $l_1 * l_2$ (the lengths of the two lines). This is why the angle between two not-normalized lines can be computed according to[4]

$$\cos(\theta) = \frac{L_1}{|L_1|} \cdot \frac{L_2}{|L_2|} \qquad (7.11)$$

or

$$\cos(\theta) = \frac{L_1^*}{|L_1^*|} \cdot \frac{L_2^*}{|L_2^*|}. \qquad (7.12)$$

Details about angles between subspaces can be found in [35].

7.4 DISTANCE BETWEEN A LINE AND A CIRCLE

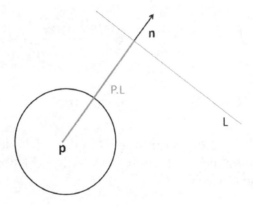

FIGURE 7.2 The inner product of a line and a circle describes the distance between the line and the center point of the circle.

The following GAALOPScript computes the inner product of the line L and the circle C.

Listing 7.5 *DistanceLineCircle.clu*: Computation of the inner product of a line and a circle.

```
1  P = createPoint(p1,p2);
2  C = P - 0.5*r*r*einf;
3  L = n1*e1+n2*e2+d*einf;
4  ?Result = L.C;
```

and results in

$$L \cdot C = n_1 p_1 + n_2 p_2 - d \qquad (7.13)$$

---

[4]Note that two lines are perpendicular to each other, if their inner product equals to zero.

$$= \mathbf{n} \cdot \mathbf{p} - d, \tag{7.14}$$

which represents the Euclidean distance between the center point of the circle and the line according to Fig. 7.2 (see Sect. 7.2 for the distance between a point and a line).

7.5 DISTANCE RELATIONS BETWEEN A POINT AND A CIRCLE

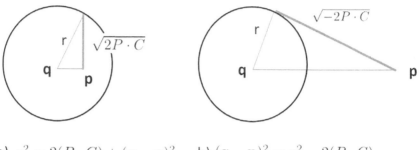

a) $r^2 = 2(P \cdot C) + (\mathbf{q} - \mathbf{p})^2$ b) $(\mathbf{q} - \mathbf{p})^2 = r^2 - 2(P \cdot C)$

FIGURE 7.3 The inner product of a point and a circle describes the distance of the bold segment depending on whether the point lies a) inside or b) outside the circle.

The following GAALOPScript computes the inner product of a point P and a circle C;

Listing 7.6 *DistancePointCircle.clu*: Computation of the inner product of a point and a circle.

```
1  P = createPoint(p1,p2);
2  Q = createPoint(q1,q2);
3  C=Q-0.5*r*r*einf;
4  Result = P.C;
```

and results in

$$P \cdot C = \frac{1}{2}r^2 - \frac{1}{2}(\mathbf{q} - \mathbf{p})^2 \tag{7.15}$$

or

$$2(P \cdot C) = r^2 - (\mathbf{q} - \mathbf{p})^2 \tag{7.16}$$

or

$$-2(P \cdot C) = (\mathbf{q} - \mathbf{p})^2 - r^2. \tag{7.17}$$

Fig. 7.3 illustrates this formula.

a) Point inside the circle: $r^2 = 2(P \cdot C) + (q - p)^2$

The triangle shown is right-angled. According to Pythagoras' theorem, the square of the radius of the circle is equal to the sum of the square of the bold segment and the square of the distance between p and q. This means that $\sqrt{2P \cdot C}$ is equal to the distance between p and the intersection of the circle with the line through p which is perpendicular to the line through p and q.

b) Point outside the circle: $(q - p)^2 = r^2 - 2(P \cdot C)$

The triangle shown is right-angled. According to Pythagoras' theorem, the square of the distance between p and q is equal to the sum of $-2P \cdot C$ and the square of the radius of the circle. This means that $\sqrt{-2P \cdot C}$ is equal to the distance between p and the tangent point to the circle.

7.6   IS A POINT INSIDE OR OUTSIDE A CIRCLE?

Looking at the square roots of the construction in Fig. 7.3 we realize

p is inside the circle $\implies P \cdot C > 0$

p is outside the circle $\implies P \cdot C < 0$

and therefore

p is on the circle $\implies P \cdot C = 0$

Now, we investigate in some more detail the inner product of a point and a circle using the Euclidean distance d according to Fig. 7.4.

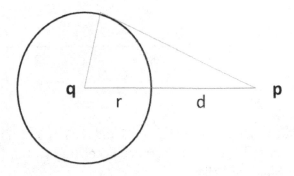

FIGURE 7.4 The inner product of a point and a circle describes the distance between the point and the tangent point of the circle according to (7.17) .

In terms of the Euclidean distance d, where

$$(d + r)^2 = (\mathbf{q} - \mathbf{p})^2 = d^2 + 2dr + r^2, \tag{7.18}$$

we get with Eq. (7.16)

$$2(P \cdot C) = r^2 - (d^2 + 2dr + r^2), \tag{7.19}$$

$$2(P \cdot C) = -d^2 - 2dr, \tag{7.20}$$

or

$$P \cdot C = I(d) = -\frac{d}{2}(d + 2r). \tag{7.21}$$

With the help of some curve sketching, we can see that this is a parabola with

$$I(0) = 0, \quad I(-2r) = 0, \tag{7.22}$$

and a maximum at

$$I(-r) = \frac{1}{2}r^2. \tag{7.23}$$

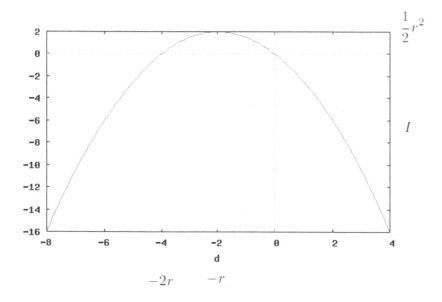

FIGURE 7.5   The inner product of a point and a circle (with radius $r = 2$) as a function of the Euclidean distance according to Eq. (7.21) (graph produced by Maxima [53]).

Fig. 7.5 shows the relation between the inner product and the Euclidean distance for a point and a circle with radius $r = 2$.

We can now see that

$\mathbf{p}$ is outside the circle $(d > 0)$ $\implies$ $I = P \cdot C < 0$

$\mathbf{p}$ is on the circle $(d = 0)$ $\implies$ $I = P \cdot C = 0$

$\mathbf{p}$ is inside the circle $(-r \leq d < 0)$ $\implies$ $0 < I = P \cdot C < \frac{1}{2}r^2$

7.7 DISTANCE TO THE HORIZON

The inner product of a point and a circle can easily be used in order to compute the distance from an observer point on the earth to the horizon (see Fig. 7.6).

According to Fig. 7.6 we can choose the coordinate frame with the center of the circle (representing the earth) at the origin e_0 and introduce a height h for the observer point in order to express its y coordinate as the sum of the radius of the earth and this height (The x coordinate in this example is 0). The GAALOPScript 7.7 computes a formula for the square of the distance to the horizon based on r and h

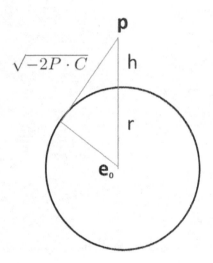

FIGURE 7.6 The inner product of a point and a circle describes the distance to the horizon.

Listing 7.7 *DistanceHorizon.clu*: Computation of the distance of an observer point to the horizon.

```
1  P = createPoint(0,r+h);
2  C=e0-0.5*r*r*einf;
3  ?DistanceSquare = -2*P.C;
```

and results in

DistanceSquare[0] $= 2.0 * h * r + h * h$. Taking concrete values for r and h leads to the following GAALOPScript

Listing 7.8 *DistanceHorizonConcrete.clu*: Computation of the distance of an observer point to the horizon with concrete values.

```
1  r = 6371;
2  h = 0.002;
3  P = createPoint(r+h,0);
4  C=e0-0.5*r*r*einf;
5  DistanceSquare = -2*P.C;
6  ?DistanceHorizon = sqrt(DistanceSquare);
```

The computation with the radius $r = 6371$ of the earth in km and $h = 0.002$ (meaning two meters) results in a distance to the horizon of 5.05 km.

7.8 DISTANCE RELATIONS BETWEEN TWO CIRCLES

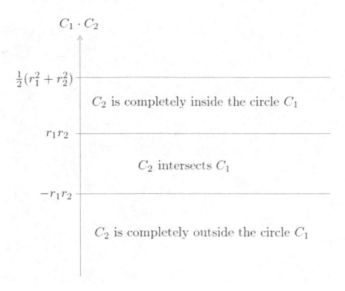

FIGURE 7.7 The geometric meaning of the inner product $C_1 \cdot C_2$ of two circles C_1 and C_2 depending on their radii r_1 and r_2.

In this section, we will see that the inner product of two circles can be used in order to describe distance relations between them [27]. Having two circles C_1, C_2 with their respective radii r_1 and r_2 (without loss of generality we assume that $r_1 \geq r_2$), we will see in this section, that the inner product between these circles has a geometric meaning according to Fig. 7.7, or in more detail:

$\frac{1}{2}(r_1^2 + r_2^2) \geq C_1 \cdot C_2 > r_1 r_2 \iff C_2$ is completely inside the circle C_1;

$C_1 \cdot C_2 = r_1 r_2 \iff C_2$ is touching the circle inside C_1;

$|C_1 \cdot C_2| < r_1 r_2 \iff C_2$ intersects C_1;

$C_1 \cdot C_2 = -r_1 r_2 \iff C_2$ is touching the circle C_1;

$C_1 \cdot C_2 < -r_1 r_2 \iff C_2$ is completely outside the circle C_1.

We develop this general result based on two examples of circles, first with equal and then with different radii.

7.8.1 Distance between Circles with Equal Radii

In Sect. 3.3.3, we already saw an example for the computation of some kind of distance measure between the circles of Fig. 7.8. Here, we investigate in more detail the inner product between two circles with equal radii, one centered at the origin and the other along the x-axis with 2D-center (x,0).

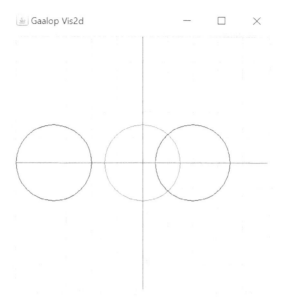

FIGURE 7.8 Visualization of *ThreeCircles.clu* (Listing 3.20): Example for the computation of the distance between two circles with equal radius according to Sect. 3.3.3.

Listing 7.9 computes the inner product $C_1 \cdot C_2(x)$ of an origin circle C_1 and a circle $C2$, both with radius r (see Fig. 7.8 with a circle at the origin, together with two examples of circles with centers along the x-axis).

Listing 7.9 *Distance TwoCircles.clu*: Computation of the inner product of two circles with equal radius.

```
1  p1=0;
2  p2=0;
3  q1=x;
4  q2=0;
5  r1=r;
6  r2=r;
7  P = createPoint(p1,p2);
8  C1 = P - 0.5*r1*r1*einf;
9  Q = createPoint(q1,q2);
10 C2 = Q - 0.5*r2*r2*einf;
11 ?I = C1.C2;
```

and results in the function for the inner product as a function of x

$$I(x) = \frac{1}{2}(2r^2 - x^2) \tag{7.24}$$

which is a parabola depending on x with

$$I(0) = \frac{1}{2}(2r^2) = r^2 \tag{7.25}$$

and the two roots

$$x_{1,2} = \pm\sqrt{2r^2} = \pm\sqrt{2}r. \tag{7.26}$$

In our example with $r = 1.5$ this is a parabola with

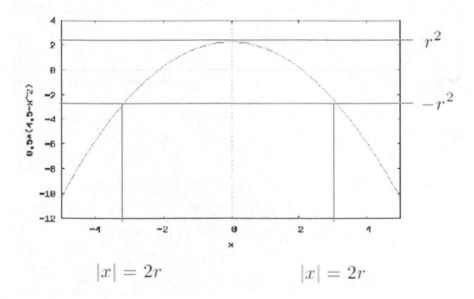

FIGURE 7.9 Function describing the inner product of two circles with equal radius r (graph produced by Maxima [53]).

$$I(0) = 2.25 \tag{7.27}$$

and the two roots

$$x_{1,2} = \pm\sqrt{4.5} \tag{7.28}$$

according to Fig. 7.9. We immediately see that the maximum of this function is reached in the case $x = 0$, where the two circles have the same center point. For tangent circles, according to Fig. 7.10, the following equation holds

$$|x| = 2r \tag{7.29}$$

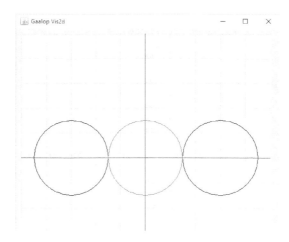

FIGURE 7.10 Tangent circles with $|x| = 2r$.

At these points, according to Eq. (7.24), the inner product of the two circles is

$$I(x) = \frac{1}{2}(2r^2 - 4r^2) \tag{7.30}$$

or

$$I(x) = \frac{1}{2}(-2r^2). \tag{7.31}$$

This means that the inner product of two tangent circles is simply

$$I(x) = -r^2. \tag{7.32}$$

The values of x for circles completely outside of each other are

$$|x| > 2r \tag{7.33}$$

or in terms of the inner product (see Fig. 7.9)

$$I(x) < -r^2. \tag{7.34}$$

The two circles are equal to each other, if the following equation holds

$$x = 0 \tag{7.35}$$

or in terms of the inner product

$$I(x) = r^2, \tag{7.36}$$

which is the maximum value of $I(x)$. In all the remaining cases

$$|I(x)| < r^2, \tag{7.37}$$

there is an intersection between the two circles. In a nutshell, we can see that

$C_1 \cdot C_2 = r^2 \iff C_2$ is equal to C_1;

$|C_1 \cdot C_2| < r^2 \iff C_2$ intersects C_1;

$C_1 \cdot C_2 = -r^2 \iff C_2$ is touching the circle C_1;

$C_1 \cdot C_2 < -r^2 \iff C_2$ is completely outside the circle C_1

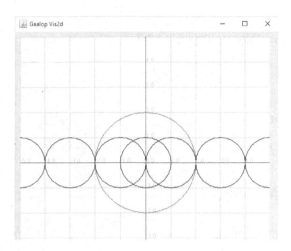

FIGURE 7.11 Example for the computation of the distance between a circle at the origin and circles with different radii along the x-axis.

7.8.2 Example of Circles with Different Radii

Let us now extend our example to the case of the different sizes of the two circles, meaning one circle can be completely inside the other. In the following, we assume that the circle C_2 is smaller than the circle C_1, which means without loss of generality we assume

$$r_1 > r_2. \tag{7.38}$$

The inner product of the two circles can be expressed according to Listing 7.10.

Listing 7.10 *Distance Two Circles2.clu*: Computation of the inner product of two circles with different radii.

```
1  p1=0;
2  p2=0;
3  q1=x;
4  q2=0;
5
6  P = createPoint(p1,p2);
```

```
7   C1 = P - 0.5*r1*r1*einf;
8   Q = createPoint(q1,q2);
9   C2 = Q - 0.5*r2*r2*einf;
10  ?I = C1.C2;
```

It computes the inner product $C_1 \cdot C_2(x)$ of the two circles and results in the function

$$I(x) = \frac{1}{2}(r_1^2 + r_2^2 - x^2) \tag{7.39}$$

which is a parabola depending on x with

$$I(0) = \frac{1}{2}(r_1^2 + r_2^2) \tag{7.40}$$

and the two roots

$$x_{1,2} = \pm\sqrt{r_1^2 + r_2^2}. \tag{7.41}$$

In our example with $r_1 = 2, r_2 = 1$ this is a parabola with

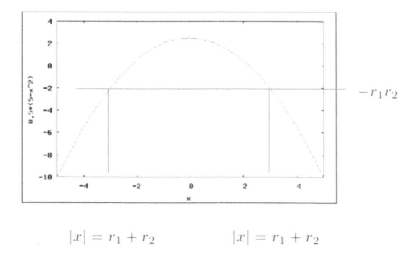

$$|x| = r_1 + r_2 \qquad\qquad |x| = r_1 + r_2$$

FIGURE 7.12 Function describing the inner product of two circles with $r_1 = 2, r_2 = 1$ according to Fig. 7.11 (graph produced by Maxima [53]).

$$I(0) = 2.5 \tag{7.42}$$

and the two roots

$$x_{1,2} = \pm\sqrt{5} \tag{7.43}$$

according to Fig. 7.12. We immediately see that the maximum of this function is reached in the case $x = 0$, where the two circles have the same center point.

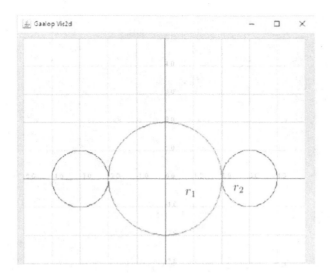

FIGURE 7.13 Tangent circles with $|x| = r_1 + r_2$.

For tangent circles according to Fig. 7.13 the following equation holds

$$|x| = r_1 + r_2. \tag{7.44}$$

At these points the inner product of the two circles is (according to Eq. (7.39))

$$I(x) = \frac{1}{2}(r_1^2 + r_2^2 - (r_1 + r_2)^2) \tag{7.45}$$

or

$$I(x) = \frac{1}{2}(r_1^2 + r_2^2 - r_1^2 - 2r_1r_2 - r_2^2) \tag{7.46}$$

or

$$I(x) = \frac{1}{2}(-2r_1r_2). \tag{7.47}$$

This means that the inner product of two tangent circles is simply

$$I(x) = -r_1r_2. \tag{7.48}$$

The values of x for circles completely outside each other are

$$|x| > r_1 + r_2 \tag{7.49}$$

or in terms of the inner product (see Fig. 7.12)

$$I(x) < -r_1r_2. \tag{7.50}$$

If a circle touches inside the other circle according to Fig. 7.14, the following

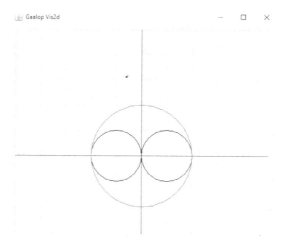

FIGURE 7.14 Inside tangent circles with $|x| = r_1 - r_2$.

equation holds

$$|x| = r_1 - r_2. \tag{7.51}$$

At these points the inner product of the two circles is

$$I(x) = \frac{1}{2}(r_1^2 + r_2^2 - (r_1 - r_2)^2) \tag{7.52}$$

or

$$I(x) = \frac{1}{2}(r_1^2 + r_2^2 - r_1^2 + 2r_1r_2 - r_2^2) \tag{7.53}$$

or

$$I(x) = \frac{1}{2}(2r_1r_2). \tag{7.54}$$

This means that the inner product of two circles touching inside is simply

$$I(x) = r_1r_2. \tag{7.55}$$

The values of x for circles completely inside of each other are

$$|x| < r_1 - r_2 \tag{7.56}$$

or in terms of the inner product

$$\frac{1}{2}(r_1^2 + r_2^2) \geq I(x) > r_1r_2. \tag{7.57}$$

In all the remaining cases

$$|I(x)| < r_1r_2, \tag{7.58}$$

there is an intersection between the two circles. In a nutshell, we can see that

$\frac{1}{2}(r_1^2 + r_2^2) \geq C_1 \cdot C_2 > r_1 r_2 \iff C_2$ is completely inside the circle C_1;

$C_1 \cdot C_2 = r_1 r_2 \iff C_2$ is touching the circle inside C_1;

$|C_1 \cdot C_2| < r_1 r_2 \iff C_2$ intersects C_1;

$C_1 \cdot C_2 = -r_1 r_2 \iff C_2$ is touching the circle C_1;

$C_1 \cdot C_2 < -r_1 r_2 \iff C_2$ is completely outside the circle C_1.

For the moment, this is the result in order to determine the meaning of the inner product of two circles for our specific example. But, we will see shortly that this result is not restricted to this case.

7.8.3 General Solution

The following GAALOPScript computes the inner product of two arbitrary circles C_1 and C_2

Listing 7.11 *Distance Two Circles3.clu*: Computation of the inner product of two arbitrary circles.

```
1  P  = createPoint(p1,p2);
2  C1 = P - 0.5*r1*r1*einf;
3  Q  = createPoint(q1,q2);
4  C2 = Q - 0.5*r2*r2*einf;
5  ?I = C1.C2;
```

and results in

$$C_1 \cdot C_2 = \frac{1}{2}(r_2^2 + r_1^2 - q_2^2 + 2 * p_2 * q_2 - q_1^2 + 2 * p_1 * q_1 - p_2^2 - p_1^2) \quad (7.59)$$

$$= \frac{1}{2}r_1^2 + \frac{1}{2}r_2^2 - \frac{1}{2}(q^2 - 2\mathbf{p} \cdot \mathbf{q} + \mathbf{p}^2) \quad (7.60)$$

$$= \frac{1}{2}(r_1^2 + r_2^2) - \frac{1}{2}(\mathbf{q} - \mathbf{p})^2. \quad (7.61)$$

We get

$$2(C_1 \cdot C_2) = r_1^2 + r_2^2 - (\mathbf{q} - \mathbf{p})^2 \quad (7.62)$$

which is very similar to Equation (7.39) of the example of Sect. 7.8.2 with different radii. Comparing the two equations, we realize that $(\mathbf{q} - \mathbf{p})^2$ easily corresponds to x^2, both describing the square of the distance between the two center points. This is why the meaning of the inner product of the example of Sect. 7.8.2 is also true for arbitrary circles with different radii. Finally, comparing the two results of Sect 7.8.1 and of Sect. 7.8.2, we realize that they can be summarized to

$\frac{1}{2}(r_1^2 + r_2^2) \geq C_1 \cdot C_2 > r_1 r_2 \iff C_2$ is completely inside the circle C_1;

$C_1 \cdot C_2 = r_1 r_2 \iff C_2$ is touching the circle inside C_1;

$|C_1 \cdot C_2| < r_1 r_2 \iff C_2$ intersects C_1;

$C_1 \cdot C_2 = -r_1 r_2 \iff C_2$ is touching the circle C_1;

$C_1 \cdot C_2 < -r_1 r_2 \iff C_2$ is completely outside the circle C_1.

since for equal radii $r = r_1 = r_2$

$r^2 \geq C_1 \cdot C_2 > r^2 \iff C_2$ is completely inside the circle C_1;

$C_1 \cdot C_2 = r^2 \iff C_2$ is touching the circle inside C_1;

$|C_1 \cdot C_2| < r^2 \iff C_2$ intersects C_1;

$C_1 \cdot C_2 = -r^2 \iff C_2$ is touching the circle C_1;

$C_1 \cdot C_2 < -r^2 \iff C_2$ is completely outside the circle C_1

and the first two cases collapse to the case

$C_1 \cdot C_2 = r^2 \iff C_2$ is equal to C_1.

Based on the general result, we are able to derive also the result for the inner product of a circle $C = C_1$ with radius $r = r_1$ and a point $P = C_2$ with radius $r_2 = 0$ to

$\frac{1}{2} r^2 \geq C \cdot P > 0 \iff C_2$ is completely inside the circle C_1;

$C \cdot P = 0 \iff P$ is touching the circle inside C;

$|C \cdot P| < 0 \iff P$ intersects C;

$C \cdot P = 0 \iff P$ is touching the circle C;

$C \cdot P < 0 \iff P$ is completely outside the circle C.

or

$\frac{1}{2} r^2 \geq C \cdot P > 0 \iff P$ is completely inside the circle C;

$C \cdot P = 0 \iff P$ is on C;

$C \cdot P < 0 \iff P$ is completely outside the circle C.

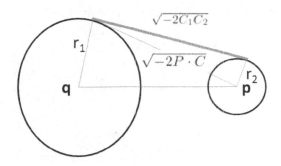

FIGURE 7.15 The bold segment describes the inner product of the two circles.

7.8.4 Geometric Meaning

Equation (7.62) means that twice the inner product of two circles is equal to the sum of the squares of their radii minus the square of the Euclidean distance between the centers of the circles. But what does this mean geometrically? The length of the bold segment in Fig. 7.15 is equal to $\sqrt{-2C_1C_2}$.

For the moment, we realize that the inner product of two circles is some kind of measure of the distance between the circles. But, what about its relation to the Euclidean distance? The Euclidean distance d between the two circles can be expressed based on the distance of the two center points and the two radii as

$$(d+r)^2 = (\mathbf{q} - \mathbf{p})^2 \tag{7.63}$$

with

$$r = r_1 + r_2. \tag{7.64}$$

With

$$(d+r)^2 = r_1^2 + r_2^2 - 2(C_1 \cdot C_2) \tag{7.65}$$

we get

$$2(C_1 \cdot C_2) = r_1^2 + r_2^2 - (d^2 + 2dr + r^2), \tag{7.66}$$

or

$$2(C_1 \cdot C_2) = r_1^2 + r_2^2 - (d^2 + 2dr + r_1^2 + 2r_1r_2 + r_2^2), \tag{7.67}$$

or

$$2(C_1 \cdot C_2) = -(d^2 + 2dr + 2r_1r_2), \tag{7.68}$$

or

$$C_1 \cdot C_2 = I(d) = -(\frac{1}{2}d^2 + dr + r_1r_2). \tag{7.69}$$

For tangent circles with $d = 0$, we immediately realize that

$$I(0) = -r_1r_2. \tag{7.70}$$

The roots of this polynomial in d can be computed as follows:

$$d_{1,2} = -r \pm \sqrt{r^2 - 2r_1 r_2} \tag{7.71}$$

or

$$d_{1,2} = \pm\sqrt{r^2 - 2r_1 r_2} - r. \tag{7.72}$$

With Eq. (7.64) we get

$$d_{1,2} = \pm\sqrt{r_1^2 + 2r_1 r_2 + r_2^2 - 2r_1 r_2} - r_1 - r_2 \tag{7.73}$$

or

$$d_{1,2} = \pm\sqrt{r_1^2 + r_2^2} - r_1 - r_2. \tag{7.74}$$

But, what is the meaning of these roots? The answer is given by the book [8]: in this case ($C_1 \cdot C_2 = 0$) the circles intersect orthogonally. In this sense, the inner product is not only a measure of distance but also a measure of orthogonality.

Note: Also in the case of two lines (see Sect. 7.3) the inner product is a measure of orthogonality.

CHAPTER 8

Transformations of Objects in Compass Ruler Algebra

Transformations of geometric objects can be easily described within Compass Ruler Algebra according to Table 8.1.

TABLE 8.1 The description of transformations of a geometric object o in Compass Ruler Algebra.

Transformation	Operator	Usage
Reflection	Line $L = \mathbf{n} + d e_\infty$	$o_L = -LoL$
Rotation	Rotor $R = \cos\left(\frac{\phi}{2}\right) - \sin\left(\frac{\phi}{2}\right) e_1 \wedge e_2$	$o_R = Ro\tilde{R}$
Translation	Translator $T = 1 - \frac{1}{2} t e_\infty$	$o_T = To\tilde{T}$
Rigid Body Motion	Motor $M = \cos\left(\frac{\phi}{2}\right) - \sin\left(\frac{\phi}{2}\right) (P \wedge e_\infty)^*$	$o_M = Mo\tilde{M}$

The reflection is a very basic operation in Geometric Algebra and transformations such as rotation and translation can be built on it. The reflection of a circle C at a line L, for instance, can be computed based on the (geometric) product $-LCL$. Rotations or translations can be described based on algebraic expressions called rotors R and translators T. The **rotor**

$$R = \cos\left(\frac{\phi}{2}\right) - \sin\left(\frac{\phi}{2}\right)(e_1 \wedge e_2), \tag{8.1}$$

describes a rotation around the origin with angle ϕ. A translation can be computed based on the **translator**

$$T = 1 - \frac{1}{2}te_\infty \tag{8.2}$$

with **t** being the 2D translation vector $t_1e_1 + t_2e_2$. While a rotor describes a rotation around the origin, the **motor**

$$M = \cos\left(\frac{\phi}{2}\right) - \sin\left(\frac{\phi}{2}\right)(P \wedge e_\infty)^* \tag{8.3}$$

is more general and describes a rotation around the point P.[1]

While the reflection operator is a vector, the operator for translations, rotations and rigid body motions is a specific bivector called **versor**. This kind of transformation of an object o can be done with the help of the following geometric product:

$$o_{transformed} = Vo\tilde{V}, \tag{8.4}$$

where V is a versor and $\tilde{V}$ is its reverse (In the reverse of a multivector the blades are with reversed order of their outer product components; for instance the reverse of $1 + e_1 \wedge e_2$ is equal to $1 + e_2 \wedge e_1$ or $1 - e_1 \wedge e_2$).

In this chapter we will learn details about these versors and how the transformations and their operators can be built-up by reflections.

8.1 REFLECTION AT THE COORDINATE AXES

A reflection of a circle at the x-axis and the y-axis can be computed with the following Listing 8.1[2]:

Listing 8.1 *ReflectCircleE1.clu*: Script for the reflection of the circle o at the line $L1 = e_1$ as well as at the line $L2 = e_2$.

```
1  o = createPoint(x,y)-0.5*r*r*einf;
2  L1 = e1;
3  L2 = e2;
4  ?o1Refl = - L1 * o * L1;
5  ?o2Refl = - L2 * o * L2;
```

---

[1]Please notice that in 3D a rigid body motion is more general in the sense that it consists of a rotation around an arbitrary line in space together with a translation in the direction of this line.

[2]Please notice regarding the lines 4 and 5, that while for the geometric product no specific symbol is used, in GAALOP "*" is needed as symbol.

The result is

$$o1Refl_1 = -x$$
$$o1Refl_2 = y$$
$$o1Refl_3 = \frac{y*y}{2} + \frac{x*x}{2} - \frac{r*r}{2}$$
$$o1Refl_4 = 1$$
$$o2Refl_1 = x$$
$$o2Refl_2 = -y$$
$$o2Refl_3 = \frac{y*y}{2} + \frac{x*x}{2} - \frac{r*r}{2}$$
$$o2Refl_4 = 1$$

or

$$o_{1Refl} = -xe_1 + ye_2 + \frac{1}{2}(x^2 + y^2 - r^2)e_\infty + e_0, \qquad (8.5)$$

which is the circle with a negated x-coordinate, meaning the circle is reflected at the y-axis,
and

$$o_{2Refl} = xe_1 - ye_2 + \frac{1}{2}(x^2 + y^2 - r^2)e_\infty + e_0, \qquad (8.6)$$

which is the circle with a negated y-coordinate, meaning the circle is reflected at the x-axis, as expected.

8.2 THE ROLE OF $E_1 \wedge E_2$

In the previous section, we realized how a reflection(at the coordinate axes) of geometric objects can be expressed with the help of the sandwich product $-LoL$. We will see in this example, that a rotation by 180 degrees can be realized based on two reflections with respect to two lines with an angle of 90 degrees between them and that the role of $e_1 \wedge e_2$ is the role of the operator for this rotation.

While in the previous listing the two reflections are performed independently, the following GAALOPScript performs them subsequently.

Listing 8.2 *CircleTwoReflections.clu*: Script for the reflection of the circle C at the lines $L1 = e_1$ and $L2 = e_2$.

```
1  C = createPoint(x,y)-0.5*r*r*einf;
2  L1=e1;
3  o1Refl = -L1 * C * L1;
4  L2=e2;
5  ?o2Refl = -L2 * o1Refl * L2;
```

Its result is

$$o2Refl_1 = -x$$
$$o2Refl_2 = -y$$
$$o2Refl_3 = \frac{y*y}{2} + \frac{x*x}{2} - \frac{r*r}{2}$$
$$o2Refl_4 = 1$$

or

$$o_{2Refl} = -xe_1 - ye_2 + \frac{1}{2}(x^2 + y^2 - r^2)e_\infty + e_0 \tag{8.7}$$

which is the circle rotated by 180 degrees.

Taking the two reflections together we can also compute the result for an arbitrary object o in one step as

$$o_{Refl2} = -L_2(-L_1oL_1)L_2 = \underbrace{(L_2L_1)}_{R}o\underbrace{(L_1L_2)}_{\tilde{R}} \tag{8.8}$$

which in our above example is

$$R = e_1e_2 = e_1 \cdot e_2 + e_1 \wedge e_2 = e_1 \wedge e_2 \tag{8.9}$$

and its reverse

$$\tilde{R} = e_2e_1 = e_2 \cdot e_1 + e_2 \wedge e_1 = e_2 \wedge e_1 = -e_1 \wedge e_2. \tag{8.10}$$

This means that there are strong relations between rotations in Compass Ruler Algebra and complex numbers. A complex number identified by the imaginary unit $i = e_1 \wedge e_2$ represents a rotation of 180 degrees around the origin. Please notice that $e_1 \wedge e_2$ also describes a point pair of the origin and infinity according to Sect. 6.2. In a nutshell, we see that $e_1 \wedge e_2$ can be visualized as the origin point as well as represent the operator for a rotation around the origin.

8.3 ARBITRARY REFLECTIONS

The following GAALOPScript computes the reflection of the circle o at an arbitrary line through the origin,

Listing 8.3 *ArbitraryReflections.clu*: Script for the reflection of the circle o at an arbitrary line through the origin.

```
1  C = createPoint (x,y) -0.5*r*r*einf;
2  L1=n1*e1+n2*e2;
3  ?o1Refl = -L1 * C * L1;
```

resulting in

$$o1Refl_1 = n2 * n2 - n1 * n1 * x - 2 * n1 * n2 * y$$
$$o1Refl_2 = n1 * n1 - n2 * n2 * y - 2 * n1 * n2 * x$$
$$o1Refl_3 = \frac{n2 * n2}{2} + \frac{n1 * n1}{2} * y * y + \frac{n2 * n2}{2} + \frac{n1 * n1}{2} * x * x...$$
$$o1Refl_4 = n2 * n2 + n1 * n1$$

The reader is encouraged to show that this describes the result of the reflection of a circle at an arbitrary line. Details about reflections in Geometric Algebra can be found in [34].

8.4   ROTOR BASED ON REFLECTIONS

It is well known in mathematics that two consecutive reflections result in a rotation by twice the angle between the two lines of reflection. If we take two arbitrary normalized lines through the origin,

$$L_1 = n_1 e_1 + n_2 e_2 \tag{8.11}$$

$$L_2 = m_1 e_1 + m_2 e_2 \tag{8.12}$$

the reverse of the rotation operator can be computed according to Sect. 8.2 as

$$\tilde{R} = L_1 L_2 = (n_1 e_1 + n_2 e_2)(m_1 e_1 + m_2 e_2) \tag{8.13}$$

or

$$\tilde{R} = (n_1 e_1 + n_2 e_2) \cdot (m_1 e_1 + m_2 e_2) + (n_1 e_1 + n_2 e_2) \wedge (m_1 e_1 + m_2 e_2). \tag{8.14}$$

According to Sect. 4.2, $\tilde{R}$ can also be written as

$$\tilde{R} = \cos(\theta) + e_1 \wedge e_2 \sin(\theta) \tag{8.15}$$

and the rotor R as its reverse

$$R = \cos(\theta) - e_1 \wedge e_2 \sin(\theta) \tag{8.16}$$

or with $\theta = \frac{\phi}{2}$

$$R = \cos\left(\frac{\phi}{2}\right) - e_1 \wedge e_2 \sin\left(\frac{\phi}{2}\right). \tag{8.17}$$

Based on this operator, the rotation of a geometric object o is performed with the help of the operation

$$o_{rotated} = Ro\tilde{R}. \tag{8.18}$$

We will show as follows that the operator R of Eq. 8.17 is equivalent to the operator

$$R = e^{-\frac{\phi}{2}e_1 \wedge e_2} \tag{8.19}$$

also describing a **rotor** for a rotation around the origin with the rotation angle ϕ.

With the help of a Taylor series, we can write

$$R = 1 + \frac{-e_1 \wedge e_2 \frac{\phi}{2}}{1!} + \frac{(-e_1 \wedge e_2 \frac{\phi}{2})^2}{2!} + \frac{(-e_1 \wedge e_2 \frac{\phi}{2})^3}{3!} + \frac{(-e_1 \wedge e_2 \frac{\phi}{2})^4}{4!} + \cdots$$

or

$$R = 1 - \frac{e_1 \wedge e_2 \frac{\phi}{2}}{1!} + \frac{(e_1 \wedge e_2 \frac{\phi}{2})^2}{2!} - \frac{(e_1 \wedge e_2 \frac{\phi}{2})^3}{3!} + \frac{(e_1 \wedge e_2 \frac{\phi}{2})^4}{4!} + \cdots$$

or, according to $i^2 = (e_1 \wedge e_2)^2 = e_1 e_2 \underbrace{e_1 e_2}_{-e_2 e_1} = -e_1 \underbrace{e_2 e_2}_{1} e_1 = -\underbrace{e_1 e_1}_{1} = -1,$

$$R = 1 - \frac{(\frac{\phi}{2})^2}{2!} + \frac{(\frac{\phi}{2})^4}{4!} - \frac{(\frac{\phi}{2})^6}{6!} + \cdots - e_1 \wedge e_2 \frac{\frac{\phi}{2}}{1!} + e_1 \wedge e_2 \frac{(\frac{\phi}{2})^3}{3!} - e_1 \wedge e_2 \frac{(\frac{\phi}{2})^5}{5!} + \cdots,$$

and therefore

$$R = \cos\left(\frac{\phi}{2}\right) - e_1 \wedge e_2 \sin\left(\frac{\phi}{2}\right). \tag{8.20}$$

8.5 TRANSLATION

In Compass Ruler Algebra, a translation can be expressed in a multiplicative way with the help of a **translator** T defined by

$$T = e^{-\frac{1}{2}t e_\infty}, \tag{8.21}$$

where $\mathbf{t}$ is a vector

$$\mathbf{t} = t_1 e_1 + t_2 e_2. \tag{8.22}$$

Application of the Taylor series

$$T = e^{-\frac{1}{2}t e_\infty} = 1 + \frac{-\frac{1}{2}t e_\infty}{1!} + \frac{(-\frac{1}{2}t e_\infty)^2}{2!} + \frac{(-\frac{1}{2}t e_\infty)^3}{3!} + \cdots \tag{8.23}$$

and the property $(e_\infty)^2 = 0$ results in the translator

$$T = 1 - \frac{1}{2}t e_\infty. \tag{8.24}$$

8.6   RIGID BODY MOTION

In Compass Ruler Algebra, a rigid body motion is a general rotation, including both a rotation and a translation as described by

$$M = TR\tilde{T}, \tag{8.25}$$

where R is a rotor, T is a translator and M is the resulting motor. A rigid body motion of an object o is described by

$$o_M = Mo\tilde{M}. \tag{8.26}$$

The following GAALOPScript

Listing 8.4   *computeMotor.clu*: Computation of a general rotation.

```
1  R =  r1  -  r2*  (e1^e2);
2  T =  1-0.5*(t1*e1+t2*e2)*einf;
3  ?M  =  T*R*  ~T;
```

results in

$$M_0 = r1$$
$$M_5 = -r2$$
$$M_6 = -r2 * t2$$
$$M_8 = r2 * t1$$

or

$$M = r_1 - r_2 e_1 \wedge e_2 - r_2 t_2 e_1 e_\infty + r_2 t_1 e_2 e_\infty \tag{8.27}$$

or

$$M = r_1 - r_2(e_1 \wedge e_2 - t_2 e_1 e_\infty + t_1 e_2 e_\infty) \tag{8.28}$$

or according to Eq. (6.6)

$$M = r_1 - r_2(P \wedge e_\infty)^* \tag{8.29}$$

with P being the conformal point of the 2D-Point (t_1, t_2).

Since r_1 and r_2 are the parameters of a rotations, this can be written in the form

$$M = \cos\left(\frac{\phi}{2}\right) - \sin\left(\frac{\phi}{2}\right)(P \wedge e_\infty)^* \tag{8.30}$$

or

$$M = \cos\left(\frac{\phi}{2}\right) - \sin\left(\frac{\phi}{2}\right) L \tag{8.31}$$

with L as the **point of rotation**

$$L = (P \wedge e_\infty)^* \tag{8.32}$$

8.7 MULTIVECTOR EXPONENTIALS

For transformations, often exponentials of multivectors are needed. They can be handled with GAALOP with a general solution based on power series expansion or, if available, with a closed-form solution for specific multivectors. Listing 8.5 computes the exponential of a motion bivector MB.

Listing 8.5 *MultivectorExponential.clu*: The exponential of a motion bivector using power series approximation and a closed form solution.

```
1  ExpApprox = {
2     1 + _P(1) + _P(1)*_P(1)/2 + _P(1)*_P(1)*_P(1)/6 +
3  _P(1)*_P(1)*_P(1)*_P(1)/24
4  }
5
6  Expon = {
7     (cos(_P(1)) + sin(_P(1))* (e1^e2))
8     + sin(_P(1))/_P(1)*_P(2)*einf
9  };
10
11 phi=0.3;
12 t1 = -3;
13 t2 = -3;
14
15 t=t1*e1 + t2*e2;
16 MB = phi*(e1^e2) + t^einf;
17 ?M1 = ExpApprox(MB);
18 ?M2 = Expon(phi,t);
```

The macro *ExpApprox* uses the first four terms of the corresponding power series. For the computation of a motor, there is also a closed-form solution by Wareham [68] available. Its 2D form is

$$M = e^B = \cos\phi + \sin\phi e_{12} + \frac{\sin\phi}{\phi}\mathbf{t}e_\infty \qquad (8.33)$$

with the motion bivector

$$B = \phi e_{12} + \mathbf{t}e_\infty, \qquad (8.34)$$

the rotation angle ϕ and the translation vector $\mathbf{t}$. This is computed by the macro *Expon*. The output of *MultivectorExponential.clu* according to the following listing

Listing 8.6 Output of *MultivectorExponential.clu*: The exponential of a motion bivector using power series approximation and a closed form solution.

```
1         M1[0] = 0.9553375; // 1.0
2         M1[5] = 0.2955; // e1 ^ e2
```

```
3    M1[6]  =  -2.955;  // e1 ^ einf
4    M1[8]  =  -2.955;  // e2 ^ einf
5    M2[0]  =  0.955336489125606;  // 1.0
6    M2[5]  =  0.2955202066613395;  // e1 ^ e2
7    M2[6]  =  -2.955202066613396;  // e1 ^ einf
8    M2[8]  =  -2.955202066613396;  // e2 ^ einf
```

shows that the motors computed by these macros are more or less the same. Computing the general output results in the following listing

Listing 8.7 Symbolic Output of *MultivectorExponential.clu*: The exponential of a motion bivector using power series approximation and a closed form solution.

```
1  M1[0]  =  0.041666666666666666 * pow(phi,4.0)
2     - (phi * phi) / 2.0 + 1.0;  // 1.0
3  M1[5]  =  phi-0.1666666666666667*phi*phi*phi;//e1^e2
4  M1[6]  =  (1.0 -0.1666666666666667*phi*phi)*t1;//e1^einf
5  M1[8]  =  (1.0 -0.1666666666666667*phi*phi)*t2;//e2^einf
6  M2[0]  =  cos(phi);  // 1.0
7  M2[5]  =  sin(phi);  // e1 ^ e2
8  M2[6]  =  (sin(phi) * t1) / phi;  // e1 ^ einf
9  M2[8]  =  (sin(phi) * t2) / phi;  // e2 ^ einf
```

8.8 INVERSION AND THE CENTER OF A CIRCLE OR POINT PAIR

Inversions are reflections not at lines but at circles. We saw in Sect. 3.4.1.3 that geometric objects resulting from inversions of lines and circles at a circle C are circles. If the objects to be inverted at a circle C move away towards infinity, the resulting circle seems to converge to the center point of C, which means it seems that the center of a circle can be computed based on the sandwich product

$$P = Ce_\infty C. \tag{8.35}$$

describing the inversion of infinity at the circle C. We can show that with the following GAALOPScript

Listing 8.8 *CircleCenterProof.clu*: Computation of the center point of a circle.

```
1  P = createPoint(p1,p2);
2  Circle = P-0.5*r*r*einf;
3  ?PC = Circle*einf*Circle;
```

resulting in the point

$$P_C = -2\left(\mathbf{p} + \frac{1}{2}\mathbf{p}^2 e_\infty + e_0\right) \tag{8.36}$$

with a homogeneous scaling factor of -2.

This sandwich product can also be used to obtain the centers of point pairs. Note, that point pairs are specific lower-dimensional circles in Compass Ruler Algebra. The center of a point pair can be computed from

$$P = P_p e_\infty P_p. \tag{8.37}$$

Applications

Robot Kinematics Using GAALOP

CONTENTS

This chapter deals with a robot kinematics application of moving a simple robot from an initial position (along the x-axis) to a target position.

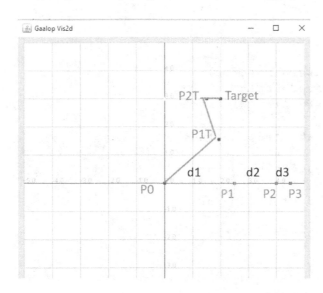

FIGURE 9.1 Inverse Kinematics of a simple robot: compute the target configuration (P0, P1T, P2T, Target) based on the initial configuration (P0, P1, P2, P3).

9.1   INVERSE KINEMATICS USING GAALOP

The simple robot of this chapter is a 3-DOF (degrees of freedom) robot acting in the plane with three links and one gripper symbolized in Fig. 9.1. Its initial position is represented by the points P0, P1, P2 and P3 with the link distances d1, d2 and d3.

The following GAALOPScript computes the joint positions in order for the robot to reach its target position under the restriction that the gripper (represented by the points P2 and P3) should at the target position be parallel to the initial pose.

Listing 9.1 *Kinematics.clu* Part I: Inverse Kinematics computations.

```
 1 d1 =2.5;
 2 d2 =1.5;
 3 d3 =0.5;
 4 tx =2;
 5 ty =3;
 6
 7 DissectFirst = {
 8   -(-sqrt( abs(_P(1)._P(1)) )+_P(1))/(einf._P(1))
 9 }
10 DissectSecond = {
11   -(sqrt( abs(_P(1)._P(1)) )+_P(1))/(einf._P(1))
12 }
13
14 P0 = e0;
15 P1 = createPoint(d1,0);
16 P2 = createPoint(d1+d2,0);
17 P3 = createPoint(d1+d2+d3,0);
18
19 Target = createPoint(tx,ty);
20
21 P2T = createPoint(tx-d3,ty);
22 C1 = P0-0.5*d1*d1*einf;
23 C2 = P2T-0.5*d2*d2*einf;
24 PP=C1^C2;
25 P1T = DissectFirst(*PP);
26
27 :Red;
28 :P0;
29 :P1;
30 :P2;
31 :P3;
32
33 :Yellow;
34 :C1;
```

```
35  : C2 ;
36
37  : Blue ;
38  : Target ;
39  : P2T ;
40  : P1T ;
41  : P0 ;
```

After the computation of the initial pose (P0, P1, P2, P3), the target position *Target* is defined based on the 2D coordinates (tx,ty). Based on this position the gripper point P2T has to be in the distance d3 from the target and parallel to the e_1-axis. This means that its 2D coordinates are (tx-d3,ty). We still have to compute the point P1T. We know that it is in a distance of d1 from the point P0 and in a distance of d2 from the point P2T. This is why we compute the circles C1 and C2 accordingly and intersect them. There are two intersecting points represented by the point pair PP. We use the macro *DissectFirst* with the formula according to Sect. 5.11 in order to extract one point for P1T.

The numerical output of Listing 9.1 is presented in the following listing,

Listing 9.2 Numerical output of *Kinematics.clu*.

```
1   P0(4)  = 1.0 // e0
2   P1(1)  = 2.5 // e1
3   P1(3)  = 3.125 // einf
4   P1(4)  = 1.0 // e0
5   P2(1)  = 4.0 // e1
6   P2(3)  = 8.0 // einf
7   P2(4)  = 1.0 // e0
8   P3(1)  = 4.5 // e1
9   P3(3)  = 10.125 // einf
10  P3(4)  = 1.0 // e0
11  Target(1)  = 2.0 // e1
12  Target(2)  = 3.0 // e2
13  Target(3)  = 6.5 // einf
14  Target(4)  = 1.0 // e0
15  P2T(1)  = 1.5 // e1
16  P2T(2)  = 3.0 // e2
17  P2T(3)  = 5.625 // einf
18  P2T(4)  = 1.0 // e0
19  P1T(1)  = 1.947019049130191 // e1
20  P1T(2)  = 1.568157142101571 // e2
21  P1T(3)  = 3.125 // einf
22  P1T(4)  = 1.0 // e0
```

with all the concrete values of our example for the initial configuration (P0, P1, P2, P3) as well as the target configuration (P0, P1T, P2T, Target). We immediately see that the e0-components of all the points are 1.0, which means all points are normalized and the e1- and e2-components correspond to the 2D coordinates of the points (a missing e1- or e2-component means that the corresponding coefficient is zero).

The 2D coordinates of the points are P0(0, 0), P1(2.5, 0), P2(4, 0), P3(4.5, 0), Target(2, 3), P2T(1.5, 3) and P1T(1.95, 1.57).

9.2 STEPS TO REACH THE TARGET

In order to reach the target position, the following GAALOPScript first computes the joint angles and then the motors (see Sect. 8.6) of each joint.

Listing 9.3 *Kinematics.clu* Part II: Motor computations in order to reach the target position.

```
1  // compute the motors for all the joints
2  L0 = e2;
3  L1 = *(P0^P1T^einf);
4  CosPhi1 = L0.L1/abs(L1);
5  Phi1 = acos(-CosPhi1);
6  M1 = cos(Phi1/2) - Sin(Phi1/2)*e1^e2;
7
8  L2 = *(P1T^P2T^einf);
9  CosPhi2 = L1.L2/(abs(L1)*abs(L2));
10 Phi2 = acos(CosPhi2);
11 M2 = cos(Phi2/2) - sin(Phi2/2)*(*(P1^einf));
12
13 L3 = *(P2T^Target^einf);
14 CosPhi3 = L2.L3/(abs(L2)*abs(L3));
15 Phi3 = acos(-CosPhi3)-3.14;
16 M3 = cos(Phi3/2) - sin(Phi3/2)*(*(P2^einf));
```

The joint angles can be computed as angles between lines according to Sect. 7.3. At the position P0 we take the line L0 perpendicular to the e2 basis vector and the line L1 through the points P0 and P1T in order to compute the cosine of the angle Phi1. For the angle Phi2 we take L1 together with the line L2 through P1T and P2T as well as L2 and the line L3 through P2T and Target for Phi3. Please note that you should be careful with the sign of the angle as well as with the decision of which angle between the two lines you choose. Now, each motor M1, M2 and M3 can be computed according to equation (8.30) in dependence of the angles Phi1, Phi2, Phi3 and the points P0=e0, P1, P2 of the initial pose of the robot.

Now, having computed all the relevant motors M1, M2 and M3, we are able to reach the target step by step.

Listing 9.4 *Kinematics.clu* Part III: Transformation computations in order to reach the target position.

```
1
2  // move step by step
3
4  // step 1
5  P31 = M3*P3*~M3;
6
7  // step 2
8  P32 = M2*P31*~M2;
9  P22 = M2*P2*~M2;
10
11  // step 3
12  P33 = M1*P32*~M1;
13  P23 = M1*P22*~M1;
14  P13 = M1*P1*~M1;
```

First, we transform the gripper link (from P2 to P3) with the motor M3. M3 as a rotation around the point P2 does not change P2, but rotates the point P3 to the rotated point P31 according to the Fig. 9.2. The robot, now, consists of the joint points P0, P1, P2 and P31.

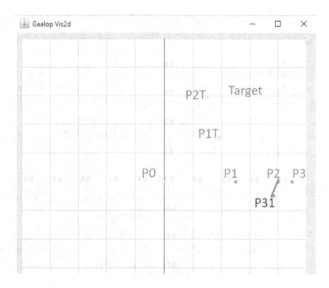

FIGURE 9.2 Step 1 to reach the target (Point P3 moved around point P2 to point P31).

In the next step, the robot with the joint points P0, P1, P2 and P31 has to be transformed by the motor M2 which is a rotation around the point P1.

When we apply this motor to the points P2 and P31, the resulting points are the points P22 and P32 according to the Fig. 9.3. The robot, now, consists of the joint points P0, P1, P22 and P32.

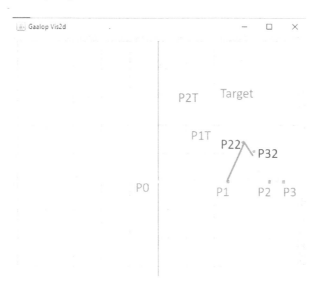

FIGURE 9.3 Step 2 to reach the target (Point P2 moved to point P22 and P3 to P32).

In the last step, the robot with the joint points P0, P1, P22 and P32 has to be transformed by the motor M1 which is a rotation around the point P0. When we apply this motor to the points P1, P22 and P32, the resulting points are the points P13, P23 and P33, which are equal to the points at the target. The robot, now, consists of the joint points P0, P1T, P2T and *Target* (according to Fig. 9.1).

9.3   MOVEMENT TOWARD THE TARGET

We just moved the robot with the help of the motors M1, M2 and M3 describing rotations based on the angles Phi1, Phi2 and Phi3. Now, we would like to perform a movement based on motors continuously changing the angles until the target is reached. This can be done based on the following GAALOPScript with a time parameter $t \in [0 .. 0.5]$.

Listing 9.5 *ContinuousMovement.clu*: Movement computations in order to reach the target position in dependence of a parameter $0 < t < 0.5$.

```
1
2  t =0.25;
3
```

```
4   // compute the motors for all the joints
5   L0 = e2;
6   L1 = *(P0^P1T^einf);
7   CosPhi1 = -L0.L1/abs(L1);
8   Phi1 = acos(CosPhi1);
9   M1 = cos(Phi1*t) - sin(Phi1*t)*e1^e2;
10
11  L2 = *(P1T^P2T^einf);
12  CosPhi2 = L1.L2/(abs(L1)*abs(L2));
13  Phi2 = acos(CosPhi2);
14  M2 = cos(Phi2*t) - sin(Phi2*t)*(*(P1^einf));
15
16  L3 = *(P2T^Target^einf);
17  CosPhi3 = -L2.L3/(abs(L2)*abs(L3));
18  Phi3 = acos(CosPhi3)-3.14;
19  M3 = cos(Phi3*t) - sin(Phi3*t)*(*(P2^einf));
20
21  // move step by step
22
23  // step 1
24  P31 = M3*P3*~M3;
25
26  // step 2
27  P32 = M2*P31*~M2;
28  P22 = M2*P2*~M2;
29
30  // step 3
31  ?P33 = M1*P32*~M1;
32  ?P23 = M1*P22*~M1;
33  ?P13 = M1*P1*~M1;
34
35  :Red;
36  :P0;
37  :P1;
38  :P2;
39  :P3;
40
41  :Green;
42  :Target;
43  :P2T;
44  :P1T;
45  :P0;
46
47  :Magenta;
48  :P33;
```

```
49  : P23 ;
50  : P13 ;
```

Figures 9.4, 9.5 and 9.6 show this movement for the parameters 0.05, 0.25 and 0.45.

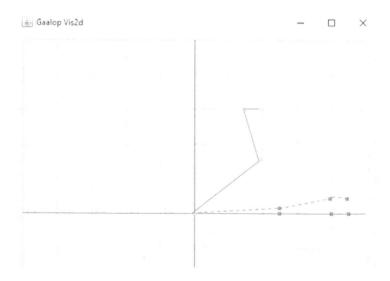

FIGURE 9.4 Movement for t =0.05.

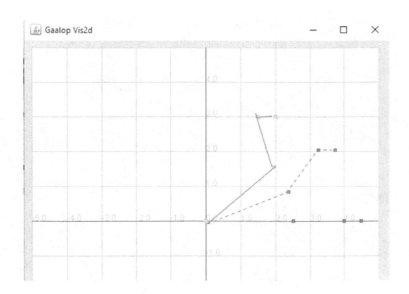

FIGURE 9.5 Movement for t =0.25.

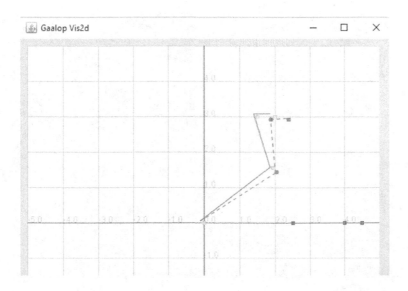

FIGURE 9.6 Movement for t =0.45.

CHAPTER 10

Detection of Circles and Lines in Images Using GAALOP

CONTENTS

This chapter presents an application dealing with the detection of circles and lines in images. We present the GAALOP implementation of the main Geometric Algebra part of the CGAVS (Conformal Geometric Algebra Voting Scheme) algorithm of the paper [65] where you can find the complete algorithm.

10.1 CGAVS ALGORITHM

The basis for the detection algorithm is an edge image showing only the discontinuities of a photograph[1]. See [61] for an edge detector based on Geometric Algebra.

The detection algorithm consists of two main stages, the local and the global voting stage. The local voting stage can be described based on Geometric Algebra. Its principle is shown in Fig. 10.1. It selects one pixel of the edge image denoted by p_0 and computes a list of pixels $\hat{P}$ in its neighborhood (the pixels $p_1 \,..\, p_4$ in Fig. 10.1). The next step is the computation of a list of lines $\hat{L}$ in the middle of two points, namely the lines between point p_0 and each of the pixels of $\hat{P}$. In the case of all pixels lying on one circle, all these lines intersect in one point, the center point of the circle. This is why we intersect each pair of lines in the next step in order to compute this circle.

---

[1]see as an example the Canny edge detector applied to a color photograph of a steam engine on Wikipedia, source: https://en.wikipedia.org/wiki/Canny_edge_detector

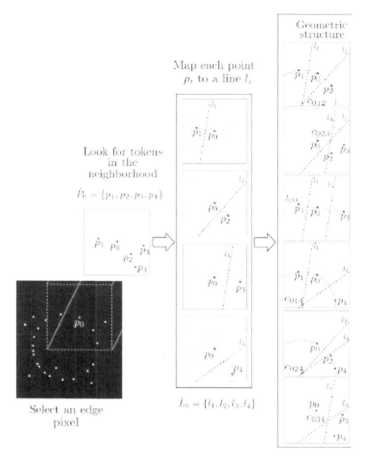

FIGURE 10.1 Local voting stage (source: [65]).

10.2   GAALOP IMPLEMENTATION

In Chapt. 4.2 of [65], the standard CGAVS implementation is described for an FPGA implementation (the same formulae can be used for implementation on other computing architectures). The following GAALOPScript computes one part of the local voting stage of the CGAVS algorithm: the two lines in the middle of the point p0 (at the origin) and the two pixels p1 and p2 as well as their intersection (the center of the circle going through p0, p1 and p2).

Listing 10.1 *EntitiesExtraction.clu*: Script for the geometric entities extraction.

```
1  px1 = 1;
2  py1 =1;
3  px2 = -3;
```

```
4    py2=-1;
5
6    ?p0 = e0;      // p0 at the origin
7    ?p1 = createPoint(px1,py1);    // Eq. (3.6)
8    ?p2 = createPoint(px2,py2);    // Eq. (3.6)
9    ?len1 = sqrt(px1*px1+py1*py1);
10   ?l1=-(px1/len1)*e1-(py1/len1)*e2-0.5*len1*einf;
11                      //Eq.(3.7)
12   //l1 = p1-p0;    // line in the middle of p1 and p0
13   ?len2 = sqrt(px2*px2+py2*py2);
14   ?l2=-(px2/len2)*e1-(py2/len2)*e2-0.5*len2*einf;
15                      //Eq.(3.7)
16   //l2 = p2-p0;    // line in the middle of p2 and p0
17   ?PpOPNS = l1^l2;                    // Eq. (4.2)
18   ?Pp=*PpOPNS;                        // Eq. (4.4)
19
20   ?IP = Pp.e0;
21
22   ?scale=-IP.einf;
23   ?c012 = IP/scale;
24
25   :p0;
26   :p1;
27   :p2;
28   :Green;
29   :l1;
30   :l2;
31   :Yellow;
32   :Pp;
33   :Blue;
34   :c012;
```

First of all, the three points p0, p1, p2 are defined according to Eq. (3.6) of [65]: p0 at the origin and p1, p2 in this example at the 2D-locations (1,1) and (-3,-1). The lines l1 and l2 are computed according to Eq. (3.7) of [65]. According to Fig. 10.2, it computes the line through p0 and pi based on its normal (the normalized 2D vector from p0 to pi) and the distance to the origin (half the distance from p0 to pi).

The intersection of the two lines is a point pair according to Eq. (4.2) of [65] and its dual a point pair according to Eq. (4.2) of [65]. Since Sect. 6.3 shows that this operation results in a point pair as the outer product of the real intersection point and infinity, we are able to compute this point IP with the help of the inner product with e_0. But, is the result really a point? When we visualize (see Fig. 10.2), we see that IP is not only the intersection point but the circle through the points p0, p1 and p2 with the center at the intersection of the two lines. Please find a general proof in Sect. 16.4.

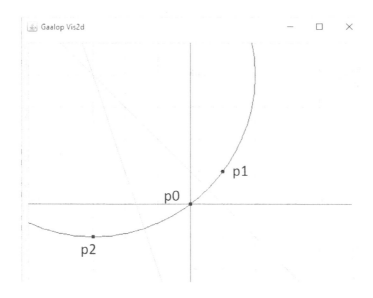

FIGURE 10.2 Visualization of *EntitiesExtraction.clu*: compute the circle through three points.

This can also be seen in the numerical output of this example according to Listing 10.2.

Listing 10.2 Numerical output of *EntitiesExtraction.clu*:

```
 1  p0(4) = 1.0 // e0
 2  p1(1) = 1.0 // e1
 3  p1(2) = 1.0 // e2
 4  p1(3) = 1.0 // einf
 5  p1(4) = 1.0 // e0
 6  p2(1) = -3.0 // e1
 7  p2(2) = -1.0 // e2
 8  p2(3) = 5.0 // einf
 9  p2(4) = 1.0 // e0
10  len1(0) = 1.414213562373095 // 1.0
11  l1(1) = -0.7071067811865475 // e1
12  l1(2) = -0.7071067811865475 // e2
13  l1(3) = -0.7071067811865476 // einf
14  len2(0) = 3.16227766016838 // 1.0
15  l2(1) = 0.9486832980505137 // e1
16  l2(2) = 0.3162277660168379 // e2
17  l2(3) = -1.58113883008419 // einf
18  PpOPNS(5) = 0.4472135954999578 // e1 ^ e2
19  PpOPNS(6) = 1.788854381999832 // e1 ^ einf
```

```
20  PpOPNS(8) = 1.341640786499874 // e2 ^ einf
21  Pp(6) = 1.341640786499874 // e1 ^ einf
22  Pp(8) = -1.788854381999832 // e2 ^ einf
23  Pp(10) = 0.4472135954999578 // einf ^ e0
24  IP(1) = -1.341640786499874 // e1
25  IP(2) = 1.788854381999832 // e2
26  IP(4) = 0.4472135954999578 // e0
27  scale(0) = 0.4472135954999578 // 1.0
28  c012(1) = -3.000000000000001 // e1
29  c012(2) = 4.000000000000002 // e2
30  c012(4) = 1.0 // e0
```

The computed point pair Pp consists of 3 components ($e_1 \wedge e_\infty$, $e_1 \wedge e_\infty$, $e_\infty \wedge e_0$), although general bivectors consist of 6 components (see Table 2.2). We realize that Pp can be written as the outer product of a vector and e_∞. The inner product with e_0 results in this vector IP with components of e_1, e_2, e_0 (being a point or circle). Its e_0-component is not 1. This is why it has to be normalized. This can be done based on the division by the e_0-component (computed by the multiplication with $-e_\infty$).

Visibility Application in 2D Using GAALOP

This chapter describes an application in 2D which is easily expandable to a 3D computer graphics application: computing the visibility of bounded spheres related to a view cone (see Sect. 15.7). In this chapter, we lay the foundations in 2D, computing the visibility of bounded circles compared to a 2D view cone.

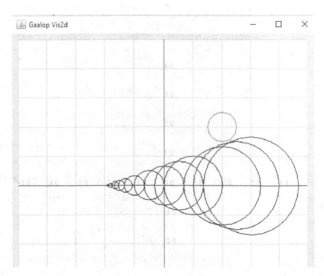

FIGURE 11.1   Is a circle outside a 2D cone modeled by circles?

11.1   IS A CIRCLE OUTSIDE A 2D CONE?

We realized in Sect. 7.8, that the inner product of two circles is able to describe a distance measure between these circles. For practical reasons, we do not take the inner product directly, but the following expression

$$d = 2(C_1 \cdot C_2 + r_1 r_2). \tag{11.1}$$

This implies

$d < 0 \implies$ one circle is completely outside the other circle

$d >= 0 \implies$ the circles are intersecting or one circle is completely inside the other circle.

Based on this observation it is easy to compute whether a circle is completely outside a 2D cone modeled by circles as visualized in Fig. 11.1. The following GAALOPScript computes a constraint for that.

Listing 11.1 *OutsideCircle2DCone.clu*: Computation of the inner product of two circles.

```
1  r=t*r1;
2  px = p1x + t*(p2x-p1x);
3  py = p1y + t*(p2y-p1y);
4  P = createPoint(px,py);
5  C1 = P - 0.5*r*r*einf;
6
7  Q = createPoint(q1,q2);
8  C2 = Q - 0.5*r2*r2*einf;
9  ?Outside = 2*(C1.C2 + r*r2);
10 //if (Outside < 0) then C2 is outside of C1
```

In dependence of $t \in [0..1]$ each circle C1 is computed. The radius r for the circles is 0 for $t = 0$ and r1 for $t = 1$. The 2D center points (px,py) of the circles interpolate between (p1x,p1y) for $t = 0$ and (p2x,p2y) for $t = 1$. The condition for C2 being outside C1 is that the inner product of the circles plus the product of the radii is smaller than 0. The variable Outside is C1.C2 + r*r2 multiplied by 2, because the result gets simpler. Now we only have to check whether Outside is smaller than 0 for all $t \in [0..1]$.

   The result of GAALOP is Outside[0] = ((((((r1 * r1 - p2y * p2y + 2.0 * p1y * p2y) - p2x * p2x + 2.0 * p1x * p2x) - p1y * p1y - p1x * p1x) **t * t** + ((2.0 * r1 * r2 + (2.0 * p2y - 2.0 * p1y) * q2 + (2.0 * p2x - 2.0 * p1x) * q1) - 2.0 * p1y * p2y - 2.0 * p1x * p2x + 2.0 * p1y * p1y + 2.0 * p1x * p1x) **t** + r2 * r2) - q2 * q2 + 2.0 * p1y * q2) - q1 * q1 + 2.0 * p1x * q1) - p1y * p1y - p1x * p1x;

which is a polynomial in t (dependent on the variables r1, r2, p1x, p1y, p2x and p2y). The only thing we have to check now is whether this polynomial is smaller than 0 for all the values of $t \in [0..1]$. If yes, the circle C2 is outside all the C1 circles.

One possibility is to compute the maximum of the polynomial. If this maximum is smaller than 0, we can be sure that all the values are smaller than 0.

The first derivative of the polynomial (produced by Maxima [53]) is

$$2 * (r1^2 - p2y^2 + 2.0 * p1y * p2y - p2x^2 + 2.0 * p1x * p2x - p1y^2 - p1x^2) * t$$

$$+2.0 * r1 * r2 + (2.0 * p2y - 2.0 * p1y) * q2 + (2.0 * p2x - 2.0 * p1x) * q1 - 2.0 * p1y * p2y$$

$$-2.0 * p1x * p2x + 2.0 * p1y^2 + 2.0 * p1x^2$$

and the second

$$2 * (r1^2 - p2y^2 + 2.0 * p1y * p2y - p2x^2 + 2.0 * p1x * p2x - p1y^2 - p1x^2).$$

Based on this information we are now able to detect whether all the values of the polynom are smaller than 0 in the interval $t \in [0..1]$ indicating that the circle C2 is outside the 2D cone modeled by the C1 circles.

11.2 VISIBILITY SEQUENCE

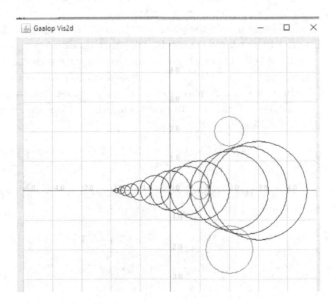

FIGURE 11.2 What is the visibility sequence of the (red) circles compared to the 2D cone modeled by circles?

The inner product approach of this chapter can also be used in order to compute a sorting sequence for bounding circles related to the view cone. For all values of the parameter t from the interval [0,1] we compute a distance measure between the bounding circle and one circle of the 2D cone based on the inner product according to the following GAALOPScript:

Listing 11.2 *DistanceCircle2DCone.clu*: Computation of the inner product of two circles.

```
1  r=t*r1;
2  px = p1x + t*(p2x-p1x);
3  py = p1y + t*(p2y-p1y);
4  P = createPoint(px,py);
5  C1 = P - 0.5*r*r*einf;
6
7  Q = createPoint(q1,q2);
8  C2 = Q - 0.5*r2*r2*einf;
9  ?Distance = 2*(C1.C2);
```

The result is the following distance function

Distance(t) = ((((((r1 * r1 - p2y * p2y + 2.0 * p1y * p2y) - p2x * p2x + 2.0 * p1x * p2x) - p1y * p1y - p1x * p1x) *** t * t** + (((2.0 * p2y - 2.0 * p1y) * q2 + (2.0 * p2x - 2.0 * p1x) * q1) - 2.0 * p1y * p2y - 2.0 * p1x * p2x + 2.0 * p1y * p1y + 2.0 * p1x * p1x) *** t** + r2 * r2) - q2 * q2 + 2.0 * p1y * q2) - q1 * q1 + 2.0 * p1x * q1) - p1y * p1y - p1x * p1x

which is a polynomial in t (dependent on the maximum radius r1 of the 2D cone, the radius r2 of the bounding circle, the starting point p1x, p1y and the end point p2x, p2y of the 2D cone). How can we compute a distance measure of the bounding circle to the 2D view cone based on this distance function?

One possibility is to compute some kind of average distance from the bounding circles to the circles of the view cone based on the integral over the interval [0,1] resulting in

```
3*r2^2+r1^2-3*q2^2+(3*p2y+3*p1y)*q2-3*q1^2+(3*p2x+3*p1x)*q1
-p2y^2-p1y*p2y-p2x^2-p1x*p2x-p1y^2-p1x^2
```

Another possibility is to compute the extremum of the above distance function. Computing the first derivative results in

```
2*(r1^2-p2y^2+2.0*p1y*p2y-p2x^2+2.0*p1x*p2x-p1y^2-p1x^2)*t
+(2.0*p2y-2.0*p1y)*q2+(2.0*p2x-2.0*p1x)*q1
-2.0*p1y*p2y-2.0*p1x*p2x+2.0*p1y^2+2.0*p1x^2
```

The extremum is reached for t = - p/q with

```
p=(p2y-p1y)*q2+(p2x-p1x)*q1-p1y*p2y-p1x*p2x+p1y^2+p1x^2
```

and

```
q=r1^2-p2y^2+2*p1y*p2y-p2x^2+2*p1x*p2x-p1y^2-p1x^2
```

Runtime-Performance Using GAALOP

CONTENTS

The main goal of GAALOP is to combine the elegance of Geometric Algebra algorithms with high runtime-performance of its implementations. This is done essentially based on the optimization of Geometric Algebra products as well as complete statements. But, there is still some optimization potential on the algorithmic level of GAALOPScripts by avoiding of

- normalizations of geometric objects

- explicitly computing the result of each statement.

For our considerations about runtime-performance we use the CGAVS algorithm of Chapt. 10.

12.1   C CODE OF THE STANDARD CGAVS IMPLEMENTATION

Let us first look at the result of the optimization process of the standard implementation of the CGAVS algorithm of Chapt. 10. For that, we take the algorithm of Listing 10.1 and ignore the statements for the visualization of Fig. 10.2 (according to Listing 12.1).

Listing 12.1   *EntitiesExtractionStandardCode.clu*: Script for the intersection of two lines l1 and l2.

```
1  ?p0 = e0;    // p0 at the origin
2  ?p1 = createPoint(px1,py1);    // Eq. (3.6)
```

```
3  ?p2 = createPoint(px2,py2);    // Eq. (3.6)
4  ?len1 = sqrt(px1*px1+py1*py1);
5  ?l1=-(px1/len1)*e1-(py1/len1)*e2-0.5*len1*einf;
6                                  //Eq.(3.7)
7  //l1 = p1-p0;   // line in the middle of p1 and p0
8  ?len2 = sqrt(px2*px2+py2*py2);
9  ?l2=-(px2/len2)*e1-(py2/len2)*e2-0.5*len2*einf;
10                                 //Eq.(3.7)
11 //l2 = p2-p0;   // line in the middle of p2 and p0
12 ?PpOPNS = l1^l2;                // Eq. (4.2)
13 ?Pp=*PpOPNS;                    // Eq. (4.4)
14
15 ?IP = Pp.e0;
16
17 ?scale=-IP.einf;
18 ?c012 = IP/scale;
```

We then generate the corresponding C code according to Listing 12.2.

Listing 12.2 EntitiesExtractionStandardCode.c: generated code of *EntitiesExtractionStandardCode.clu.*

```
1  void calculate(float px1, float px2, float py1,
2  float py2, float c012[16], float IP[16], float l1[16],
3  float l2[16], float len1[16], float len2[16],
4  float p0[16], float p1[16], float p2[16],
5  float Pp[16], float PpOPNS[16], float scale[16]) {
6
7  p0[4] = 1.0; // e0
8  p1[1] = px1; // e1
9  p1[2] = py1; // e2
10 p1[3] = (py1 * py1) / 2.0 + (px1 * px1) / 2.0; // einf
11 p1[4] = 1.0; // e0
12 p2[1] = px2; // e1
13 p2[2] = py2; // e2
14 p2[3] = (py2 * py2) / 2.0 + (px2 * px2) / 2.0; // einf
15 p2[4] = 1.0; // e0
16 len1[0] = sqrtf(py1 * py1 + px1 * px1); // 1.0
17 l1[1] = (-px1) / len1[0]; // e1
18 l1[2] = (-py1) / len1[0]; // e2
19 l1[3] = -0.5 * len1[0]; // einf
20 len2[0] = sqrtf(py2 * py2 + px2 * px2); // 1.0
21 l2[1] = (-px2) / len2[0]; // e1
22 l2[2] = (-py2) / len2[0]; // e2
23 l2[3] = -0.5 * len2[0]; // einf
24 PpOPNS[5] = l1[1] * l2[2] - l1[2] * l2[1]; // e1 ^ e2
25 PpOPNS[6] = l1[1] * l2[3] - l1[3] * l2[1]; // e1 ^ einf
```

```
26  PpOPNS[8] = 11[2] * 12[3] - 11[3] * 12[2]; //e2 ^ einf
27  Pp[6] = PpOPNS[8]; // e1 ^ einf
28  Pp[8] = (-PpOPNS[6]); // e2 ^ einf
29  Pp[10] = PpOPNS[5]; // einf ^ e0
30  IP[1] = (-Pp[6]); // e1
31  IP[2] = (-Pp[8]); // e2
32  IP[4] = Pp[10]; // e0
33  scale[0] = IP[4]; // 1.0
34  c012[1] = IP[1] / scale[0]; // e1
35  c012[2] = IP[2] / scale[0]; // e2
36  c012[4] = IP[4] / scale[0]; // e0
37  }
```

First, the representation of the points p0, p1 and p2 is computed based on the point coordinates px1, py1 and px2, py2. These are the input values of the algorithm and all the other computations are intermediate results leading to the final result c012. We recognize that GAALOP computes only the coefficients of multivectors which are explicitly needed.

12.2 AVOIDING NORMALIZATIONS

Looking at the generated C code of Listing 12.2 and its most expensive operations, we notice the sqrt and division operations for the computation of the lines l1 and l2 as well as the division for the circle c012. All these operations are needed for some normalization of these geometric objects. The following Listing 12.3 computes the same visualization according to Fig. 10.2 but it does not need the normalizations.

Listing 12.3 *EntitiesExtractionAvoidingNormalizations.clu*: Script for the intersection of two lines l1 and l2.

```
1   px1 = 1;
2   py1 =1;
3   px2 = -3;
4   py2=-1;
5
6   ?p0 = e0;    // p0 at the origin
7   ?p1 = createPoint(px1,py1);    // Eq. (3.6)
8   ?p2 = createPoint(px2,py2);    // Eq. (3.6)
9   ?l1 = p1-p0;    // line in the middle of p1 and p0
10  ?l2 = p2-p0;    // line in the middle of p2 and p0
11  ?PpOPNS = l1^l2;                // Eq. (4.2)
12  ?Pp=*PpOPNS;                    // Eq. (4.4)
13
14  ?IP = Pp.e0;
15
```

```
16  : p0 ;
17  : p1 ;
18  : p2 ;
19  : Green ;
20  : l1 ;
21  : l2 ;
22  : Yellow ;
23  : Pp ;
24  : Blue ;
25  : IP ;
```

Here we compute the resulting (not normalized) geometric object IP which is visualized completely in the same way as before the normalized c012. This shows that it is not always needed to normalize geometric objects. Especially within an algorithm normalization computations can be avoided. For instance outer products of not normalized geometric objects result in geometric objects which are again not normalized but correct with respect to its geometric meaning. In Listing 12.1 the two lines to be intersected are normalized (their normal vectors have a length of 1). In Listing 12.3 we use the nice feature of CGA that the difference of two points results in the line(2D)/plane(3D) between the two points (see Sect. 3.2.6). This operation is not leading to normalized objects. We can see that in the following numerical output listing 12.4

Listing 12.4   Numerical output of *EntitiesExtractionAvoidingNormalizations.clu*:

```
1   p0(4)  =  1.0  // e0
2   p1(1)  =  1.0  // e1
3   p1(2)  =  1.0  // e2
4   p1(3)  =  1.0  // einf
5   p1(4)  =  1.0  // e0
6   p2(1)  = -3.0  // e1
7   p2(2)  = -1.0  // e2
8   p2(3)  =  5.0  // einf
9   p2(4)  =  1.0  // e0
10  l1(1)  =  1.0  // e1
11  l1(2)  =  1.0  // e2
12  l1(3)  =  1.0  // einf
13  l2(1)  = -3.0  // e1
14  l2(2)  = -1.0  // e2
15  l2(3)  =  5.0  // einf
16  PpOPNS(5)  =  2.0  // e1 ^ e2
17  PpOPNS(6)  =  8.0  // e1 ^ einf
18  PpOPNS(8)  =  6.0  // e2 ^ einf
19  Pp(6)  =  6.0  // e1 ^ einf
20  Pp(8)  = -8.0  // e2 ^ einf
```

```
21  Pp(10) = 2.0 // einf ^ e0
22  IP(1) = -6.0 // e1
23  IP(2) = 8.0 // e2
24  IP(4) = 2.0 // e0
```

l1 consists of the 2D normal vector (1,1) and l2 of the normal vector (-3,-1) which both do not have a length of 1. Nevertheless, looking at the output of IP and comparing it with the result of IP and its normalized form c012 in Listing 10.2 we realize that they are describing the same geometric object.

Looking now at the generated C code of Listing 12.5

Listing 12.5 EntitiesExtractionAvoidingNormalizations.c: generated code of the middle part of *EntitiesExtractionAvoidingNormalizations.clu*

```
1   void calculate(float px1, float px2, float py1,
2   float py2, float IP[16], float l1[16], float l2[16],
3   float p0[16], float p1[16], float p2[16],
4   float Pp[16], float PpOPNS[16])
5   {
6     p0[4] = 1.0; // e0
7     p1[1] = px1; // e1
8     p1[2] = py1; // e2
9     p1[3] = (py1 * py1)/2.0 + (px1 * px1)/2.0; // einf
10    p1[4] = 1.0; // e0
11    p2[1] = px2; // e1
12    p2[2] = py2; // e2
13    p2[3] = (py2 * py2)/2.0 + (px2 * px2)/2.0; // einf
14    p2[4] = 1.0; // e0
15    l1[1] = p1[1]; // e1
16    l1[2] = p1[2]; // e2
17    l1[3] = p1[3]; // einf
18    l2[1] = p2[1]; // e1
19    l2[2] = p2[2]; // e2
20    l2[3] = p2[3]; // einf
21    PpOPNS[5] = l1[1] * l2[2] - l1[2] * l2[1]; // e1^e2
22    PpOPNS[6] = l1[1] * l2[3] - l1[3] * l2[1]; //e1^einf
23    PpOPNS[8] = l1[2] * l2[3] - l1[3] * l2[2]; //e2^einf
24    Pp[6] = PpOPNS[8]; // e1 ^ einf
25    Pp[8] = (-PpOPNS[6]); // e2 ^ einf
26    Pp[10] = PpOPNS[5]; // einf ^ e0
27    IP[1] = (-Pp[6]); // e1
28    IP[2] = (-Pp[8]); // e2
29    IP[4] = Pp[10]; // e0
30  }
```

we realize that the computations are much easier and sqrt and division operations are no longer needed.

12.3 AVOIDING EXPLICIT STATEMENT COMPUTATIONS

As an example, we focus on the intersection calculator module of [65] and take the specific equations (4.1), (4.2) and (4.4) of the algorithm and implement them in the following GAALOPScript.

Listing 12.6 *EntitiesExtraction1.clu*: Script for the intersection of two lines l1 and l2.

```
1  ?l1 = nx1*e1 + ny1*e2 + dh1*einf;   // Eq. (4.1)
2  ?l2 = nx2*e1 + ny2*e2 + dh2*einf;   // Eq. (4.1)
3  ?PpOPNS = l1^l2;                     // Eq. (4.2)
4  ?Pp=*PpOPNS;                         // Eq. (4.4)
```

All the four statements are computed explicitly (indicated by the leading question marks) and the GAALOP result of the multivectors l1, l2 PpOPNS and Pp is:

Listing 12.7 EntitiesExtraction1.c: resulting C code of *EntitiesExtraction1.clu*.

```
1   l1[1] = nx1; // e1
2   l1[2] = ny1; // e2
3   l1[3] = dh1; // einf
4   l2[1] = nx2; // e1
5   l2[2] = ny2; // e2
6   l2[3] = dh2; // einf
7   PpOPNS[5] = l1[1] * l2[2] - l1[2] * l2[1];//e1^e2
8   PpOPNS[6] = l1[1] * l2[3] - l1[3] * l2[1];//e1^einf
9   PpOPNS[8] = l1[2] * l2[3] - l1[3] * l2[2];//e2^einf
10  Pp[6] = PpOPNS[8];    // e1^einf
11  Pp[8] = (-PpOPNS[6]);// e2^einf
12  Pp[10] = PpOPNS[5];   // einf^e0
```

The result of PpOPNS is exactly the result of equation (4.3) of [65] and Pp describes the result of equation (4.5).

In order to compute only the final multivector Pp, the GAALOPScript 12.6 has to be changed to

Listing 12.8 *EntitiesExtraction1a.clu*: Script for the geometric entities extraction.

```
1  l1 = nx1*e1 + ny1*e2 + dh1*einf;   // Eq. (4.1)
2  l2 = nx2*e1 + ny2*e2 + dh2*einf;   // Eq. (4.1)
3  PpOPNS = l1^l2;                     // Eq. (4.2)
4  ?Pp=*PpOPNS;                        // Eq. (4.4)
```

with only one question mark for Pp resulting in the simpler result:

Listing 12.9 EntitiesExtraction1a.c: resulting C-code of *EntitiesExtraction1a.clu*.

```
1 | Pp[6]  = dh2 * ny1 - dh1 * ny2; // e1 ^ einf
2 | Pp[8]  = dh1 * nx2 - dh2 * nx1; // e2 ^ einf
3 | Pp[10] = nx1 * ny2 - nx2 * ny1; // einf ^ e0
```

making everything in one step.

If we extend the GAALOPScript according to

Listing 12.10 *EntitiesExtraction2a.clu*: Script for the geometric entities extraction.

```
1 | p0 = e0;    // p0 at the origin
2 | p1 = createPoint(px1,py1);    // Eq. (3.6)
3 | p2 = createPoint(px2,py2);    // Eq. (3.6)
4 | l1 = p1-p0;   // line in the middle of p1 and p0
5 | l2 = p2-p0;   // line in the middle of p2 and p0
6 | PpOPNS = l1^l2;                 // Eq. (4.2)
7 | Pp=*PpOPNS;                     // Eq. (4.4)
8 | ?IP = Pp.e0;
```

we compute the resulting multivector IP directly based on the points p0, p1, p2 in one step. GAALOP computes the resulting point pair multivector as

Listing 12.11 EntitiesExtraction2a.c: resulting C-code of *EntitiesExtraction2a.clu*.

```
1 | IP[1] = -0.5 *py1*py2*py2 + ((py1*py1)/2.0
2 |    + (px1*px1)/2.0)*py2-(px2 * px2)/2.0*py1; //e1
3 | IP[2] = px1/2.0*py2*py2 - px2/2.0*py1*py1
4 |    + px1/2.0*px2*px2 - (px1*px1)/2.0*px2; // e2
5 | IP[4] = px1 * py2 - px2 * py1; // e0
```

To avoid the divisions, we also can multiply IP by -2 (which does not change the geometric object) according to the following GAALOPScript

Listing 12.12 *EntitiesExtraction2b.clu*: Script for the geometric entities extraction.

```
1 | p0 = e0;    // p0 at the origin
2 | p1 = createPoint(px1,py1);    // Eq. (3.6)
3 | p2 = createPoint(px2,py2);    // Eq. (3.6)
4 | l1 = p1-p0;   // line in the middle of p1 and p0
5 | l2 = p2-p0;   // line in the middle of p2 and p0
6 | PpOPNS = l1^l2;                 // Eq. (4.2)
7 | Pp=*PpOPNS;                     // Eq. (4.4)
8 | ?IP = -2*Pp.e0;
```

and the result of the complete computation from the points via the lines to the point pair is simply

Listing 12.13 EntitiesExtraction2b.c: resulting C-code of *EntitiesExtraction2b.clu.*

```
1  IP[1] = py1 * py2 * py2
2      + (-(py1 * py1) - px1 * px1) * py2
3      + px2 * px2 * py1; // e1
4  IP[2] = -px1 * py2 * py2 + px2 * py1 * py1
5      - px1 * px2 * px2 + px1 * px1 * px2; // e2
6  IP[4] = 2.0 * px2 * py1 - 2.0 * px1 * py2; // e0
```

with only the point coordinates px1, py1 and px2, py2 as input values. The big advantage of this solution compared to Listing 12.2 is that

- the square roots and divisions are no longer needed,

- instead of 11 multivectors only 1 has to be computed,

- instead of 30 assignments to multivector coefficients only 3 are needed.

12.4 NEW CGAVS ALGORITHM

The Listing 10.1 presents an implementation of the standard CGAVS algorithm close to standard geometry consisting of operations such as the intersection of lines. Thinking about the core of the algorithm we realize that we have to compute the circle going through three points. How can we easily express this in Compass Ruler Algebra? From Table 5.1 we know that in order to compute a circle going through three points, we have the possibility to compute the dual of the outer product of them. This is done by the following GAALOPScript:

Listing 12.14 *CircleFromPoints.clu*: Script for the computation of the circle through three points of the geometric entities extraction algorithm.

```
1  p0=e0;
2  p1 = createPoint(px1,py1);
3  p2 = createPoint(px2,py2);
4
5  C = *(p0^p1^p2);
6  ?CTimes2 = 2*C;
```

It results in the following C/C++ code:

Listing 12.15 CircleFromPoints.c: C-code for the computation of the circle through three points of the geometric entities extraction algorithm.

```
1  CTimes2[1] = py1 * py2 * py2 +
2  ((-(py1*py1)) - px1*px1)*py2 + px2*px2*py1; // c1
```

```
3   CTimes2 [2]  =  (( -(px1 * py2 * py2)) + px2 * py1 * py1)
4   - px1 * px2 * px2 + px1 * px1 * px2;  // e2
5   CTimes2 [4]  =  2.0*px2 * py1 - 2.0*px1 * py2;  // e0
```

Comparing this result with Listing 12.13 shows completely the same result, but with a much smaller GAALOPScript.

What happens if the points are co-linear? This is handled by the following GAALOPScript.

Listing 12.16 *CircleToLine.clu*: Script for the computation of the circle through three co-linear points of the geometric entities extraction algorithm.

```
1   px2= px1*t;
2   py2= py1*t;
3
4   p0=e0;
5   p1 = createPoint(px1,py1);
6   p2 = createPoint(px2,py2);
7
8   C = *(p0^p1^p2);
9   ?CTimes2 = 2*C;
```

We additionally request that there is a linear dependence between the 2D-vectors (px1, px2) and (py1, py2). It results in the following C/C++ code:

Listing 12.17 CircleToLine.c: C-code for the computation of the line through three co-linear points of the geometric entities extraction algorithm.

```
1   CTimes2 [1]  =  (py1*py1*py1 + px1*px1*py1)*t*t
2                  + ((-(py1*py1*py1)) - px1*px1*py1)*t;//e1
3   CTimes2 [2]  =  ((-(px1*py1*py1)) - px1*px1*px1)*t*t
4                  + (px1*py1*py1 + px1*px1*px1)*t;//e2
```

describing the corresponding line. Comparing the two results for circles and lines we can see that we always can take the circle computations. In the case of a line the e_0 component is zero which can be taken as a criterion whether the result is a circle or a line (does px2*py1-px1*py2 equal to zero?).

Altogether this example shows the expressive power of Geometric Algebra and its potential for highly performant implementations.

12.5   HARDWARE IMPLEMENTATION BASED ON GAALOP

There are two ways to generate hardware (HW) implementations based on GAALOP. One is to automatically generate an optimized FPGA (field programmable gate array) description according to [26] and [67] and the other one is the new *GAPPCO*[31] solution with many advantages.

GAPPCO is a recent coprocessor design combining both the advantages of optimizing software with a configurable hardware able to implement arbitrary Geometric Algebra algorithms. The idea is to have a fixed hardware, easy and fast to configure for different algorithms. Compared to standard hardware architectures it makes use of variable-size vectors, pipelining and fast register access. The GAPPCO design is based on *GAPP* (Geometric Algebra Parallelism Programs), a language describing the general structure of the computations after the GAALOP optimization process. As described in the book [26], Geometric Algebra algorithms with all kinds of products of multivectors always have the same principle structure. Please refer to [31] for the general design and for the first version called GAPPCO I.

As an example, we generate GAPP code from Listing 12.14[1] resulting in the following listing:

Listing 12.18 CircleFromPoints.gapp: GAPP-code for the computation of the circle through three points of the geometric entities extraction algorithm.

```
1  assignInputsVector inputsVector = [px1,px2,py1,py2];
2  resetMv CTimes2[16];
3  setVector ve0 = {inputsVector[2,-2,-0,1]};
4  setVector ve1 = {inputsVector[3,2,0,1]};
5  setVector ve2 = {inputsVector[3,3,3,2]};
6  dotVectors CTimes2[1] = <ve0,ve1,ve2>;
7
8  setVector ve3 = {inputsVector[-0,1,-0,0]};
9  setVector ve4 = {inputsVector[3,2,1,0]};
10 setVector ve5 = {inputsVector[3,2,1,1]};
11 dotVectors CTimes2[2] = <ve3,ve4,ve5>;
12
13 setVector ve6 = {2.0, 2.0};
14 setVector ve7 = {inputsVector[1,0]};
15 setVector ve8 = {inputsVector[2,3]};
16 dotVectors CTimes2[4] = <ve6,ve7,ve8>;
```

Note: implementing this algorithm on FPGA has the big advantage that only integer arithmetic is needed.

---

[1] In order to generate GAPP code, please select "GAPP" for Optimization and "GAPP CodeGenerator" for CodeGenerator.

CHAPTER 13

Fitting of Lines or Circles into Sets of Points

CONTENTS

One big advantage of Compass Ruler Algebra is that both circles and lines have the same algebraic structure. This is why we are now able to make the approach of computing the best-fitting line or circle into a set of points $\mathbf{p}_i \in \mathbb{R}^2$, $i \in \{1, \ldots, n\}$ (please refer to [26] for a 3D version of this algorithm).

Lines and circles are both vectors of the form

$$S = s_1 e_1 + s_2 e_2 + s_3 e_\infty + s_4 e_0. \tag{13.1}$$

Note, that a line is represented, if $s_4 = 0$.

The points are specific vectors of the form

$$P_i = \mathbf{p}_i + \frac{1}{2}\mathbf{p}_i^2 e_\infty + e_0. \tag{13.2}$$

In order to solve the fitting problem, we do the following:

Use the distance measure between a point and a circle or line with the help of the inner product.

Use a least-squares approach to minimize the squares of the distances between the points and the circle or line.

Solve the resulting eigenvalue problem.

The main benefit of this approach is that it fits either a line or a circle, depending on which one fits better.

13.1 DISTANCE MEASURE

From Sect. 7, we already know that a distance measure between a point P_i and a circle or line S can be defined with the help of their inner product

$$P_i \cdot S = \left(\mathbf{p}_i + \frac{1}{2}\mathbf{p}_i^2 e_\infty + e_0\right) \cdot (\mathbf{s} + s_3 e_\infty + s_4 e_0). \qquad (13.3)$$

The GAALOPScript

Listing 13.1 *IPPointVector.clu*: Script for the computation of a distance measure for a point and a vector representing a line or a circle.

```
1  Pi = createPoint(pi1,pi2);
2  S = s1*e1 + s2*e2+ s3*einf + s4*e0;
3  ?Result = Pi.S;
```

results in

$$Result_0 = -0.5 * pi2 * pi2 - \frac{pi1 * pi1}{2} * s4 - s3 + pi2 * s2 + pi1 * s1$$

or

$$P_i \cdot S = \mathbf{p_i} \cdot \mathbf{s} - s_3 - \frac{1}{2}s_4\mathbf{p}_i^2 \qquad (13.4)$$

or, equivalently,

$$P_i \cdot S = \sum_{j=1}^{4} w_{i,j} s_j, \qquad (13.5)$$

where

$$w_{i,j} - \begin{cases} p_{i,j}, & j \in \{1,2\} \\ -1, & j = 3 \\ -\frac{1}{2}\mathbf{p}_i^2, & j = 4. \end{cases} \qquad (13.6)$$

13.2 LEAST-SQUARES APPROACH

In the least-squares sense, we consider the minimum of the sum of the squares of the distances (expressed in terms of the inner product) between all of the points considered and the line/circle,

$$\min \sum_{i=1}^{n} (P_i \cdot S)^2. \qquad (13.7)$$

In order to obtain this minimum, it can be rewritten in bilinear form as

$$\min(s^T Bs), \qquad (13.8)$$

where

$$s^T = (s_1, s_2, s_3, s_4),$$ (13.9)

and the 4×4 matrix

$$B = \begin{pmatrix} b_{1,1} & b_{1,2} & b_{1,3} & b_{1,4} \\ b_{2,1} & b_{2,2} & b_{2,3} & b_{2,4} \\ b_{3,1} & b_{3,2} & b_{3,3} & b_{3,4} \\ b_{4,1} & b_{4,2} & b_{4,3} & b_{4,4} \end{pmatrix}$$ (13.10)

has entries

$$b_{j,k} = \sum_{i=1}^{n} w_{i,j} w_{i,k}.$$ (13.11)

The matrix B is symmetric, since $b_{j,k} = b_{k,j}$. We consider only normalized results such that $s^T s = 1$. A conventional approach to such a constrained optimization problem is to introduce

$$L = s^T B s - 0 = s^T B s - \lambda(s^T s - 1),$$ (13.12)

$$s^T s = 1,$$ (13.13)

$$B^T = B.$$ (13.14)

The necessary conditions for a minimum are

$$0 = \nabla L = 2 \cdot (Bs - \lambda s) = 0$$ (13.15)

$$\rightarrow Bs = \lambda s.$$ (13.16)

The solution of the minimization problem is given by the eigenvector of B that corresponds to the smallest eigenvalue.

Figures 13.1 and 13.2 illustrate two properties of the distance measure in this approach, dealing with the double squaring of the distance and the limiting process for the distance in the case of a line considered as a circle of infinite radius.

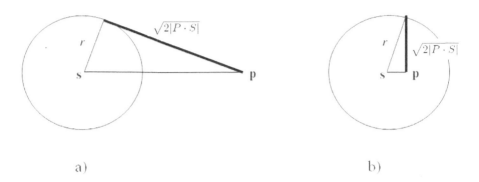

a)   b)

FIGURE 13.1 The inner product $P \cdot S$ of a point and a circle on the one hand already describes the square of a distance, but on the other hand has to be squared again in the least-squares method, since the inner product can be positive or negative depending on whether a) the point **p** lies outside the circle or b) the point **p** lies inside the circle.

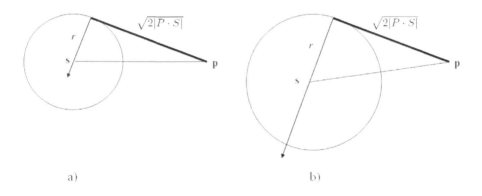

a)   b)

FIGURE 13.2 The constraint $s^T s = 1$ leads implicitly to a scaling of the distance measure such that it gets smaller with increasing radius; if the radius increases from the one in a) via the radius in b) and further to an infinite radius, the distance measure gets zero for a line considered as a circle of infinite radius.

CRA-Based Robotic Snake Control

This chapter treats robot kinematics in a mathematically advanced manner: the control of a snake robot is presented based on differential kinematics. It summarizes the results of the papers [47], [45] and [46]. In contrast to Chapt. 9, it models links based on point pairs.

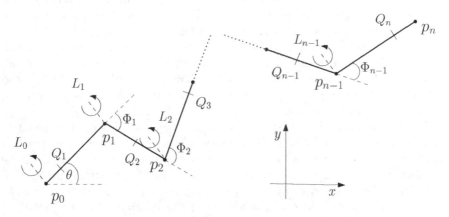

FIGURE 14.1   Snake robot model.

We consider an n-link snake robot moving on a planar surface. More precisely, it is a model when to each link, two wheels are attached and thus the possible movement directions are determined uniquely. Calculations in CRA allow the wheels to be placed not in the link's center of mass only, but their position is arbitrary within each link. The aim is to find the complete kinematic description. Although we handle only the case that the links are of a constant length 1, the generalization to an arbitrary length of each link is obvious. If the generalized coordinates are considered, the non–holonomic forward kinematic equations can be understood as a Pfaff system. In the classical approach, local controllability is discussed by means of differential geometry and Lie algebras, see [62, 54]. Our aim is to translate the whole kinematics into the language of CRA, where both linear objects and spheres of dimensions 1 and 0, see [13, 26, 57, 69], are easy to transform. The classical approach composes the kinematic chain of homogeneous matrices using the moving frame method and Euler angles [52].

In particular, the point pair is consequently used to derive the kinematic equations and for the control of the robotic snake. More precisely, to any link of a snake a single point pair is assigned and the mechanism is transformed by rotations and translations. We introduce the differential kinematic equations, as well as the non–holonomic conditions, respectively. Also the singularity conditions are formulated. The advantage of the CRA description lies in the simplicity of the model modifications, i.e. variable link length and variable wheel position. Furthermore, we fully use the advantage of CRA in operations representation; precisely rotations and translations are represented by particular CRA elements.

14.1 ROBOTIC SNAKES

The topic of the snake–like robots goes back to early 1970's when S. Hirose formulated the essential model design and developed limbless locomotors; for the complex review of his work see [33]. He started the first bio-mechanical study using real snakes and designed the first snake-like robot based on so–called serpentine locomotion. The first designs of Hirose's snake robots had modules with small passive wheels, and since then, most of the current developments by Downling (1997) [9], Chirikjian and Burdick (1990) [4], and Ostrowski (1995) [55] keep using the snake robots with wheels in order to facilitate forward propulsion.

The snake robot described in this chapter consists of n rigid links interconnected by $n-1$ motorized joints. To each line, a pair of wheels is attached to provide an important snake-like property that the ground friction in the direction perpendicular to the link is considerably higher than the friction of a simple forward move. In particular, this prevents slipping sideways. To describe the actual position of a snake robot the generalized coordinates

$$q = (x, y, \theta, \Phi_1, ..., \Phi_{n-1}) \qquad (14.1)$$

are considered, see Figure 14.1.

Note that a fixed coordinate system (x, y) is attached. For sake of simplicity, we consider the links to be of constant length 1 but the generalization to arbitrary lengths is obvious. The points $p_i := (x_i, y_i), i = 0, ..., n$, denote the endpoints of each link and by $Q_i = r_i p_i + (1 - r_i) p_{i-1}$, $r_i \in \langle 0, 1 \rangle$, $i = 1, ..., n$, we denote the points where the wheels are attached to the particular link. Then, the distance $|Q_i p_{i-1}|$ is equal to r_i. If the absolute angle of the i–th link, i.e. the angle between the link and the x–axis, is denoted by θ_i then the position of Q_i w.r.t. the global $x - y$ axes is then expressed as

$$Q_{x,i} = P_{x,0} + \sum_{j=1}^{i-1} \cos \theta_j + r_i \cos \theta_i,$$

$$Q_{y,i} = P_{y,0} + \sum_{j=1}^{i-1} \sin \theta_j + r_i \sin \theta_i. \tag{14.2}$$

Note that to recover the generalized coordinates one has to consider the assertion

$$\theta_i = \sum_{j=1}^{i-1} \Phi_j + \theta.$$

Furthermore, the linear velocity of Q_i can be determined by taking the derivative of (14.2) and thus the nonholonomic equations are obtained.

This gives us a rough idea of the snake model in the Euclidean plane. To describe the robotic snake by means of CRA we use as a central object the point pairs

$$P_i = p_{i-1} \wedge p_i, \ i = 1, ..., n$$

and thus the i–th link is represented by a point pair P_i. Anyway, if the position of a particular joint p_i is needed, one can consider the projection of a point pair onto its endpoints in the form[1]

$$p_{i-1} = \frac{-\sqrt{P_i \cdot P_i} + P_i}{e_\infty \cdot P_i}, \quad p_i = \frac{\sqrt{P_i \cdot P_i} + P_i}{e_\infty \cdot P_i}.$$

14.2 DIRECT KINEMATICS

The direct kinematics for the snake robot is obtained similarly to the kinematics for serial robot arms [69]. For the case of a 3–link robotic snake see Sect. 14.4. In general, it is given by a succession of generalized rotations R_i and translations T_i. The composition of R_i and T_i is a motor and will be denoted by M_i. Particularly, the actual position of a joint p_i at a general point $q = (x, y, \theta, \Phi_1, ..., \Phi_{n-1}) \in \mathbb{R}^{n+2}$ is computed from its initial position $p_i(0)$ by

$$p_i(q) = M_i p_i(0) \tilde{M}_i, \ i = 0, ..., n,$$

where $p_i(0)$ is the initial position of p_i and M_i is a motor defined as

---

[1] Here, we are using dual points in comparison to Sect. 5.11.

$$M_0 := T = 1 - \frac{1}{2}(xe_1 + ye_2)e_\infty,$$

i.e. T stands for the translation from the origin to the position of the head point, and

$$M_i = R_i...R_1T \text{ for } i > 0,$$

i.e. the product of rotations. These can be determined inductively as follows (see Eq. (8.31)):

$$R_{i+1} = e^{-\Phi_i L_i} = \cos\frac{\Phi_i}{2} - \sin\frac{\Phi_i}{2}L_i,$$

$$L_i = M_i L_i(0)\tilde{M}_i,$$

where L_i are the points of rotations placed in the corresponding joints[2], see Figure 14.1. Furthermore, the wheel position at the link P_i is calculated as

$$Q_i = M_i Q_i(0)\tilde{M}_i. \tag{14.3}$$

FIGURE 14.2 Snake robot initial position.

In the following, the snake robot's initial position is the one depicted in Figure 14.2, i.e $q_0 = (0, ..., 0)$. Then

$$p_j(0) = je_1 + \frac{1}{2}j^2 e_\infty + e_0, \tag{14.4}$$

and

$$L_j(0) = jc_2 \wedge c_\infty - e_1 \wedge e_2 \tag{14.5}$$

for the point of rotation[3] at each joint j according to the following GAALOP-Script based on Eq. (8.31)

Listing 14.1 *PointOfRotation.clu*: Script for the computation of the point of rotation at joint j.

```
1  px=j;
2  py=0;
3  P  =  createPoint(px,py);
4  ?Lj  =  *(P^einf);
```

This gives us the whole kinematic chain which corresponds to equations (14.2).

---

[2]The L_i can be seen as the axis of rotation in 3D perpendicular to the plane at the corresponding point of rotation.

[3]Please notice, that the point of rotation according to Eq. (14.5) corresponds exactly to the 3D computations of a line perpendicular to the x-y-plane in Sect. 15.6.

14.2.1   Singular positions

The singular positions, i.e. singular points in $\mathbb{R}^{n+2}$ in generalized coordinates, can be characterized as those positions that do not allow the snake–like motion without breaking the nonholonomic constraints. Note that the whole mechanism can move linearly (see the zero initial position in Figure 14.2) or rotate around a given point without changing the Φ_i–coordinates. Yet these motions are not considered as snake–like as they are not a consequence of the mechanism construction but rather of the outer forces.

The singular point example is the following: a position is singular if all wheel axes o_i intersect in one point, see Figure 14.3. In CRA description, this

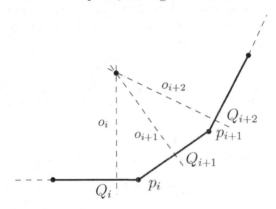

FIGURE 14.3   Snake robot's singular position.

condition is simply expressed as

$$o_i^* \wedge o_j^* \wedge o_k^* = 0 \tag{14.6}$$

for any three indices $i, j, k \in \{1, ..., n\}$. To describe the singular position in CRA completely, we add that for a wheel position Q_i on the link P_i the wheel axle is expressed as

$$o_i^* = (Q_i \wedge e_\infty) \cdot (P_i \wedge e_\infty).$$

14.3   DIFFERENTIAL KINEMATICS

The differential kinematics will be obtained by the differentiation of this kinematic chain as follows. For the wheel position point $Q_i \in P_i$ we have

$$\dot{Q}_i = Q_i \cdot (e_1 \wedge e_\infty)\dot{x} + Q_i \cdot (e_2 \wedge e_\infty)\dot{y} + \sum_{j=0}^{i-1}(Q_i \cdot L_j)\dot{\Phi}_j. \tag{14.7}$$

Note that according to [47], the equation (14.7) holds for any other point on the link P_i and, with minor modification, for any CRA object attached to P_i at the position of Q_i .

If we consider the wheels positions as a vector $Q = (Q_i)^T$, the equation (14.7) transforms as

$$\dot{Q} = J\dot{q}, \tag{14.8}$$

where

$$J = \big(\underbrace{Q_i \cdot (e_1 \wedge e_\infty) \quad Q_i \cdot (e_2 \wedge e_\infty)}_{n \times 2} \mid \underbrace{Q_i \cdot L_{j-1}}_{n \times n}\big), \ i,j = 1, ..., n,$$

plays the role of a Jacobi matrix, particularly a matrix of inner products of Q_i and axes of rotations or translations.

As the wheels do not slip to the side direction, the velocity vector must be parallel to $\dot{Q}_i$ and the constraint condition, i.e. the nonholonomic constraint, is in terms of CRA expressed as

$$\dot{Q}_i \wedge P_i \wedge e_\infty = 0. \tag{14.9}$$

Thus if we substitute (14.7) in (14.9), we obtain a system of linear ODEs

$$A\dot{q} = 0, \tag{14.10}$$

where $A = (a_{ij})$ plays the role of the Pfaff matrix and is of a simple form

$$a_{ij} = J_{ij} \wedge P_i \wedge e_\infty. \tag{14.11}$$

Note that the elements of A are just pseudoscalar multiples and thus A can be understood as a matrix over the field of functions. To specify the elements of A more precisely we formulate the following Lemma. For sake of simplicity, we suppose that each link length is 1 and the wheels attached to the i–th link are represented by a point Q_i but both these parameters can be easily generalized. In particular, each link can be of a different length and Q_i can stand for any CRA object.

Lemma 14.3.1 *If $Q_i = r_i p_i + (1 - r_i)p_{i-1}$, $r_i \in \langle 0, 1 \rangle$, is a point on the link P_i, then*

$$(Q_i \cdot L_{i-1}) \wedge P_i \wedge e_\infty = r_i I, \tag{14.12}$$

where I is a pseudoscalar.

For proof see [45].

If we denote the element $e_i \wedge e_\infty$ by $e_{i\infty}$ then Lemma 14.3.1 directly implies the following

Proposition 14.3.2 *The Pfaff matrix A of the system (14.17) is of the form*

$$A = (b|B),$$

where

$$b = \begin{pmatrix} e_{1\infty} \wedge P_1 & e_{2\infty} \wedge P_1 \\ e_{1\infty} \wedge P_2 & e_{2\infty} \wedge P_2 \\ \vdots & \vdots \\ e_{1\infty} \wedge P_n & e_{2\infty} \wedge P_n \end{pmatrix}$$

and

$$B = \begin{pmatrix} r_1 & 0 & \cdots & \cdots & 0 \\ (Q_2 \cdot L_0) \wedge P_2 \wedge e_\infty & r_2 & & & \\ \vdots & & \ddots & & 0 \\ (Q_n \cdot L_0) \wedge P_n \wedge e_\infty & \cdots & (Q_n \cdot L_{n-2}) \wedge P_n \wedge e_\infty & r_n \end{pmatrix}.$$

One can see that the matrix B in the form from Proposition 14.3.2 is a lower triangle block matrix with nonzero elements on the diagonal (and thus always invertible). The inverse of such matrix is easy to express. Thus generally, the matrix $A = (b|B)$, where $b = (b_{ij})$, $i = 1, ..., n$, $j = 1, 2$, and $B = (B_{ij})$, $i, j = 1, ..., n$, are $n \times 2$ and $n \times n$ dimensional matrices, respectively, and their elements can be expressed as

$$\begin{aligned} b_{ij} &= (Q_i \cdot e_{j\infty}) \wedge P_i \wedge e_\infty = e_{j\infty} \wedge P_i, \\ B_{ij} &= (Q_i \cdot L_{j-1}) \wedge P_i \wedge e_\infty. \end{aligned} \tag{14.13}$$

Thus if we consider the control $u = (u_1, u_2)$, $u_1 = \dot{x}$, $u_2 = \dot{y}$ then the control matrix G is in the form

$$G = \begin{pmatrix} E \\ -B^{-1}b \end{pmatrix},$$

where E is a 2×2 unit matrix, the system (14.8) can be written as

$$\dot{Q} = J\dot{q} = JGu,$$

where

$$JG = \underbrace{\left(Q \cdot e_{1\infty} \quad Q \cdot e_{2\infty} \right)}_{n \times 2} - \underbrace{\left(Q \cdot L_{j-1} \right)}_{n \times n} \underbrace{B^{-1}b}_{n \times 2}.$$

Moreover, once the model is reformulated in this sense, the velocity equations of the wheel points change accordingly, e.g. the equation of the wheel point Q_n attached to the last link P_n will be of the form

$$\dot{Q}_n = Q_n \cdot \begin{pmatrix} \ell_1 & \ell_2 \end{pmatrix} \begin{pmatrix} u_1 \\ u_2 \end{pmatrix},$$

where

$$\ell_1 = e_{1\infty} - \sum_{i=1}^{n} L_{i-1} \sum_{j=1}^{n} B_{ij}^{-1} b_{j1}$$

$$\ell_2 = e_{2\infty} - \sum_{i=1}^{n} L_{i-1} \sum_{j=1}^{n} B_{ij}^{-1} b_{j2}$$

provided that B_{ij} and b_{ij} are the elements of B and b, respectively, specified by (14.13).

14.4   3-LINK SNAKE MODEL

The snake robot described in this chapter consists of 3 rigid links of constant length 2 interconnected by motorized 2 joints. To each line, in the center of mass, a pair of wheels is attached to provide an important snake-like property that the ground friction in the direction perpendicular to the link is considerably higher than the friction of a simple forward move. In particular, this prevents slipping sideways. To describe the actual position of a snake robot we need the set of 5 generalized coordinates

$$q = (x, y, \theta, \Phi_1, \Phi_2) \tag{14.14}$$

which describe the configuration of the snake robot as shown in Figure 14.4. (for more details see [45]).

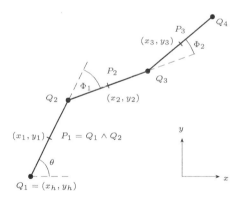

FIGURE 14.4 Snake robot model.

Note that a fixed coordinate system (x, y) is attached. The points $p_1 := (x_1, y_1)$, $p_2 := (x_2, y_2)$, $p_3 := (x_3, y_3)$ denote the centers of mass of each link. To describe the robotic snake we use as a central object the couple of point pairs

$$(P_1, P_3)$$

where $P_1 = Q_1 \wedge Q_2$ and $P_3 = Q_3 \wedge Q_4$, where Q_i are joints. Consequently, the kinematic equations can be assessed and if we consider the projections

$$Q_2 = -\frac{\sqrt{P_1 \cdot P_1} + P_1}{e_\infty \cdot P_1}, \quad Q_3 = \frac{\sqrt{P_3 \cdot P_3} + P_3}{e_\infty \cdot P_3},$$

we are able to express the first point coordinates from any point pair. Finally, denote $P_2 = Q_2 \wedge Q_3$. The coordinates of a particular position vectors are expressed as

$$p_1 = P_1 e_\infty \tilde{P}_1, \text{ s.t. } P_1 = R_\theta T_{x,y} P_{1,0} \tilde{T}_{x,y} \tilde{R}_\theta,$$
$$p_2 = P_2 e_\infty \tilde{P}_2, \text{ s.t. } P_2 = R_{\Phi_1} R_\theta T_{x,y} P_{2,0} \tilde{T}_{x,y} \tilde{R}_\theta \tilde{R}_{\Phi_1},$$
$$p_3 = P_3 e_\infty \tilde{P}_3, \text{ s.t. } P_3 = R_{\Phi_2} R_{\Phi_1} R_\theta T_{x,y} P_{3,0} \tilde{T}_{x,y} \tilde{R}_\theta \tilde{R}_{\Phi_1} \tilde{R}_{\Phi_2}$$

and, for the robotic snake initial position $x = y = \theta = \Phi_1 = \Phi_2 = 0$, three appropriate point pairs are established as

$$P_{1,0} = (e_0) \wedge (2e_1 + 2e_\infty + e_0) = 2e_0 e_1 - 2e_+ e_-,$$
$$P_{2,0} = (2e_1 + 2e_\infty + e_0) \wedge (4e_1 + 8e_\infty + e_0) = 2e_0 e_1 + 8e_1 e_\infty - 6e_+ e_-,$$
$$P_{3,0} = (4e_1 + 8e_\infty + e_0) \wedge (6e_1 + 18e_\infty + e_0) = 2e_0 e_1 + 24e_1 e_\infty - 10e_+ e_-.$$

Now, the transformations corresponding to the generalized coordinates can be written as

$$T_{x,y} = 1 - \frac{1}{2}(xe_1 + ye_2)e_\infty, \quad T_{Q_1} = 1 - \frac{1}{2}Q_1 e_\infty, \quad T_{Q_2} = 1 - \frac{1}{2}Q_2 e_\infty,$$
$$R_\theta = \cos\frac{\theta}{2} - L_0 \sin\frac{\theta}{2}, \text{ where } L_0 = T_{x,y} e_1 e_2 \tilde{T}_{x,y},$$
$$R_{\Phi_1} = \cos\frac{\Phi_1}{2} - L_1 \sin\frac{\Phi_1}{2}, \text{ where } L_1 = T_{Q_2} e_1 e_2 \tilde{T}_{Q_2},$$
$$R_{\Phi_2} = \cos\frac{\Phi_2}{2} - L_2 \sin\frac{\Phi_2}{2}, \text{ where } L_2 = T_{Q_3} e_1 e_2 \tilde{T}_{Q_3}.$$

The direct kinematics for the snake robot is obtained similarly to the kinematics for serial robot arms [69]. In general, it is given by a succession of generalized rotations R_i and it is valid for all geometric objects, including point pairs. A point pair P in a general position is computed from its initial position P_0 as follows

$$P = \prod_{i=1}^{n} R_i P_0 \prod_{i=1}^{n} \tilde{R}_{n-i+1}.$$

Unlike the fixed serial robot arms, we allow R_i to be also a translation. We view translations as degenerate rotations. Then the differential kinematics is expressed by means of the total differential as follows

$$dP = \sum_{j=1}^{n} \partial_{q_j} \left(\prod_{i=1}^{n} R_i P_0 \prod_{i=1}^{n} \tilde{R}_{n-i+1} \right) dq_j.$$

Since both the translations and the rotations can be expressed as exponentials, and $dR = d(e^{-\frac{1}{2}qL}) = -\frac{1}{2}RLdq$, the straightforward computation leads to the following assertion, [69, 47]:

$$dP = \sum_{j=1}^{n} [P \cdot L_j] dq_j.$$

If R_i is a translation, then the axis of rotation L_i is given by a linear combination of bivectors that contain e_∞. In our particular case, previous considerations lead to the following general expression of $\dot{p}_i$, where the range of the subscript i is determined by the number of links:

$$\dot{p}_i = \partial_t(P_i e_\infty \tilde{P}_i) = \dot{P}_i e_\infty \tilde{P}_i + P_i e_\infty \dot{\tilde{P}}_i$$

$$= \sum_{j=1}^{n} [P_i \cdot L_j] e_\infty \tilde{P}_i dq_j + P_i e_\infty \sum_{j=1}^{n} [\tilde{L}_j \cdot \tilde{P}_i] dq_j$$

$$= \frac{1}{2} \sum_{j=1}^{n} \left(P_i L_j e_\infty \tilde{P}_i - L_j P_i e_\infty \tilde{P}_i + P_i e_\infty \tilde{L}_j \tilde{P}_i - P_i e_\infty \tilde{P}_i \tilde{L}_j \right) dq_j$$

In the last line we used the definition of the scalar product and the fact that L_i contains only bivectors. This also implies that L_i always commutes with e_∞ (in the case of a translation the product vanishes), and that $\tilde{L}_i = -L_i$. Thus we get

$$\dot{p}_i = \frac{1}{2} \sum_{j=1}^{n} \left(-L_j P_i e_\infty \tilde{P}_i - P_i e_\infty \tilde{P}_i \tilde{L}_j \right) dq_j = \sum_{j=1}^{n} [p_i \cdot L_j] dq_j,$$

i.e. the same formula of the differential kinematics holds also for the link centers p_i. Concretely, we obtain the system

$$\dot{p}_1 = [p_1 \cdot e_1 e_\infty] \dot{x} + [p_1 \cdot e_2 e_\infty] \dot{y} + [p_1 \cdot L_0] \dot{\theta},$$
$$\dot{p}_2 = [p_2 \cdot e_1 e_\infty] \dot{x} + [p_2 \cdot e_2 e_\infty] \dot{y} + [p_2 \cdot L_0] \dot{\theta} + [p_2 \cdot L_1] \dot{\Phi}_1,$$
$$\dot{p}_3 = [p_3 \cdot e_1 e_\infty] \dot{x} + [p_3 \cdot e_2 e_\infty] \dot{y} + [p_3 \cdot L_0] \dot{\theta} + [p_3 \cdot L_1] \dot{\Phi}_1 + [p_3 \cdot L_2] \dot{\Phi}_2$$

which in the matrix notation is of the form

$$\dot{p} = J\dot{q}, \tag{14.15}$$

where q are our coordinates (14.14) and $J = (j_{kl})$ is a 3×5 matrix with the elements defined by

$$j_{i1} = [p_i \cdot e_1 e_\infty], \; j_{i2} = [p_i \cdot e_2 e_\infty],$$
$$j_{ik} = [p_i \cdot L_{k-3}] \text{ for } 3 \le k < 3 + i,$$
$$j_{ik} = 0 \text{ for } 3 + i \le k.$$

As the wheels do not slip to the side direction, the velocity constraint condition is satisfied for each link i and in terms of CGA can be written as

$$\dot{p}_i \wedge P_i \wedge e_\infty = 0. \tag{14.16}$$

Thus if we substitute (14.15) in (14.16), we obtain a system of linear ODEs, which has a simple Pfaff matrix form

$$A\dot{q} = 0, \tag{14.17}$$

where $A = (a_{ij})$ is a matrix with the elements defined by

$$a_{ik} = j_{ik} \wedge P_i \wedge e_\infty. \tag{14.18}$$

Note that the entries of A are multiples of the pseudoscalar I and hence A can be considered simply as a matrix over the field of functions. For example, the solution of this system with respect to $\dot{\theta}$ parameterized by $\dot{x}, \dot{y}$, (i.e. $\dot{x} = t_1$ and $\dot{y} = t_2$) is of the form

$$\dot{\theta} = -\frac{[p_1 \cdot e_1 e_\infty] \wedge P_1 \wedge e_\infty}{[p_1 \cdot L_0] \wedge P_1 \wedge e_\infty} t_1 - \frac{[p_1 \cdot e_2 e_\infty] \wedge P_1 \wedge e_\infty}{[p_1 \cdot L_0] \wedge P_1 \wedge e_\infty} t_2.$$

The straightforward computation leads to $[p_1 \cdot L_0] \wedge P_1 \wedge e_\infty = 2I$, i.e. the solution always exists, because $([p_1 \cdot L_0] \wedge P_1 \wedge e_\infty)^{-1} = -\frac{1}{2}I$. The system matrix is singular in case that the wheel axes, i.e. lines perpendicular to each link containing the link center point, intersect in precisely one point or are parallel, see Figure 14.2.

In our setting this is one condition only because in CRA the parallel lines intersect in exactly one point which is e_∞. It is easy to see that this happens in such case that all joints lie on a single circle; i.e. in CRA they satisfy a simple condition

$$P_1 \wedge P_3 = 0. \tag{14.19}$$

Finally, note that the non–singular solution forms a 2–dimensional distribution which can be parameterized e.g. as follows:

$$\dot{q} = G \begin{pmatrix} t_1 \\ t_2 \end{pmatrix}, \tag{14.20}$$

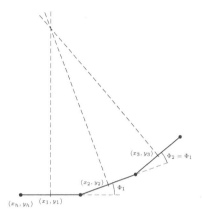

FIGURE 14.5   Snake robot singular position.

where $G = (g_{ij})$ is a 2×5 control matrix with the elements defined by

$$g_{11} = 1, \; g_{12} = 0, \; g_{21} = 0, \; g_{22} = 1, \; g_{31} = \cos(\theta), \; g_{32} = \sin(\theta),$$
$$g_{41} = -2\cos(\Phi_1)\sin(\theta) + \sin(\theta + \Phi_1) - \sin(\theta),$$
$$g_{42} = 2\cos(\Phi_1)\cos(\theta) - \cos(\theta + \Phi_1) + \cos(\theta),$$
$$g_{51} = 4\cos(\Phi_1)\cos(\Phi_2)\sin(\theta) - 2\sin(\theta + \Phi_1)\cos(\Phi_2) + 2\cos(\Phi_1)\sin(\theta)$$
$$\quad - 2\cos(\Phi_1 + \Phi_2)\sin(\theta) - \sin(\theta + \Phi_1) + \sin(\theta + \Phi_1 + \Phi_2),$$
$$g_{52} = -4\cos(\Phi_1)\cos(\Phi_2)\cos(\theta) + 2\cos(\theta + \Phi_1)\cos(\Phi_2)$$
$$\quad - 2\cos(\Phi_1)\cos(\theta) + 2\cos(\Phi_1 + \Phi_2)\cos(\theta) + \cos(\theta + \Phi_1)$$
$$\quad - \cos(\theta + \Phi_1 + \Phi_2).$$

Thus if we consider the snake robot configuration space with coordinates (14.14) as a 5–dimensional manifold M, the solution above forms a couple of vector fields g_1 and g_2.

It is clear, that the space span$\{g_1, g_2\}$ determines the set of accessible velocity vectors and thus, taking into account the vector field flows $\exp(tg_1)$, $\exp(tg_2)$, the possible trajectories of the snake robot. On the other hand, due to non–commutativity of $\exp(tg_1)$, $\exp(tg_2)$, the robot can move even along the flow of the Lie bracket by means of the composition

$$\exp(-tg_2) \circ \exp(-tg_1) \circ \exp(tg_2) \circ \exp(tg_1).$$

Extending this idea, the space Q_q of all movement directions in point q is given by taking all possible Lie brackets of $g_1(q)$ and $g_2(q)$ and the resulting vector fields. From the geometric control theory point of view, it is quite necessary that the dimension of Q_q is equal to the dimension of the tangent space $T_q M, q \in M$, which in our case is 5. Note that this is the condition on the model local controllability given by the Rashevsky–Chow Theorem. In our case, it is easy to show that in regular points q indeed

$$Q_q = \text{span}\{g_1, g_2, [g_1, g_2], [g_1, [g_1, g_2]], [g_2, [g_1, g_2]]\} \cong T_q M.$$

Thus the tangent space to the configuration space of the snake robot is equipped with a $(2, 3, 5)$ filtration.

Please refer to Sect. 15.5 for a simulation of this application using the CLUCalc software package.

Expansion to 3D Computations

This chapter presents some information about how to extend the 2D information of the previous chapters to 3D.

Throughout this book, we used Compass Ruler Algebra. According to Chapt. 2, this Geometric Algebra consists of 16 basis **blades** (combinations of outer products of e_1 and e_2 and two additional basis vectors of Compass Ruler Algebra, e_0 and e_∞)[1]. Please refer to Chapt. 5 for details about the algebraic structure of Compass Ruler Algebra, which is the **Conformal Geometric Algebra (CGA)** in 2D.

Compared to Compass Ruler Algebra, Conformal Geometric Algebra consists of one additional basis vector e_3 for 3D space. Table 15.1 lists all the 32 basis blades of CGA. The basis vectors $e_1, e_2, e_3, e_0, e_\infty$ are the five grade-1 blades of this algebra. There is one grade-0 blade (the scalar) and one grade-5 blade (the pseudoscalar). Linear combinations of the 10 grade-2 blades, the 10 grade-3 blades and the five grade-4 blades are called **bivectors, trivectors** and **quadvectors**. A linear combination of blades with different grades is called a **multivector**. Multivectors are the main algebraic elements of Conformal Geometric Algebra.

---

[1]Please refer to Table 2.2 on page 10.

TABLE 15.1   The 32 blades of the 5D Conformal Geometric Algebra (Conformal Geometric Algebra)

Grade	Term	Blades	No.
0	Scalar	1	1
1	Vector	$e_1, e_2, e_3, e_\infty, e_0$	5
2	Bivector	$e_1 \wedge e_2,\ \ e_1 \wedge e_3,\ \ e_1 \wedge e_\infty,$ $e_1 \wedge e_0,\ \ e_2 \wedge e_3,\ \ e_2 \wedge e_\infty,$ $e_2 \wedge e_0,\ \ e_3 \wedge e_\infty,\ \ e_3 \wedge e_0,$ $e_\infty \wedge e_0$	10
3	Trivector	$e_1 \wedge e_2 \wedge e_3,\ \ e_1 \wedge e_2 \wedge e_\infty,\ \ e_1 \wedge e_2 \wedge e_0,$ $e_1 \wedge e_3 \wedge e_\infty,\ \ e_1 \wedge e_3 \wedge e_0,\ \ e_1 \wedge e_\infty \wedge e_0,$ $e_2 \wedge e_3 \wedge e_\infty,\ \ e_2 \wedge e_3 \wedge e_0,\ \ e_2 \wedge e_\infty \wedge e_0,$ $e_3 \wedge e_\infty \wedge e_0$	10
4	Quadvector	$e_1 \wedge e_2 \wedge e_3 \wedge e_\infty,$ $e_1 \wedge e_2 \wedge e_3 \wedge e_0,$ $e_1 \wedge e_2 \wedge e_\infty \wedge e_0,$ $e_1 \wedge e_3 \wedge e_\infty \wedge e_0,$ $e_2 \wedge e_3 \wedge e_\infty \wedge e_0$	5
5	Pseudoscalar	$e_1 \wedge e_2 \wedge e_3 \wedge e_\infty \wedge e_0$	1

15.1   CLUCALC FOR 3D VISUALIZATIONS

In 3D, we use the *CLUCalc* software package [57] for the visualization of Geometric Algebra algorithms[2]. CLUCalc is freely available for download at [58][3]. With the help of CLUCalc, you will be able to edit and run scripts called *CLUScripts*. The screenshot in Fig. 15.1 shows the three windows of CLUCalc:

- an editor window;

- a visualization window;

- an output window.

With the help of the editor window, you can easily edit your CluScripts, in the visualization window you are able to see the 3D visualizations and in the output window numerical values of multivectors are shown. That is comparable to GAALOP and GAALOPScript. While GAALOPScript focuses on symbolic computing, CLUScript is able to

- extend Geometric Algebra algorithms with control flow and loops,

- use tools such as sliders,

---

[2]GAALOP can be used for symbolic computations in 3D (see Sect. 15.6).

[3]We recommend downloading version 4.3.3 in order to be able to run the examples of the book [26].

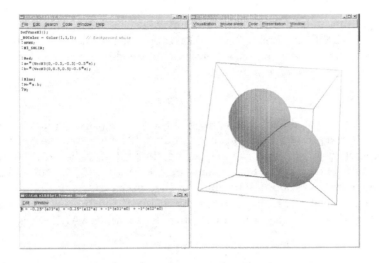

FIGURE 15.1 Screenshot of a CLUCalc algorithm for the intersection of two spheres.

- annotate visualizations (see Fig. 15.2).

15.2 THE GEOMETRIC OBJECTS OF CGA

TABLE 15.2 The two representations (IPNS and OPNS) of 3D conformal geometric entities. The IPNS and OPNS representations are dual to each other, which is indicated by the asterisk symbol.

Entity	IPNS representation	OPNS representation
Point	$P = \mathbf{x} + \frac{1}{2}\mathbf{x}^2 e_\infty + e_0$	
Sphere	$S = P - \frac{1}{2}r^2 e_\infty$	$S^* = P_1 \wedge P_2 \wedge P_3 \wedge P_4$
Plane	$\pi = \mathbf{n} + d e_\infty$	$\pi^* = P_1 \wedge P_2 \wedge P_3 \wedge e_\infty$
Circle	$C = S_1 \wedge S_2$	$C^* = P_1 \wedge P_2 \wedge P_3$
Line	$L = \pi_1 \wedge \pi_2$	$L^* = P_1 \wedge P_2 \wedge e_\infty$
Point pair	$Pp = S_1 \wedge S_2 \wedge S_3$	$Pp^* = P_1 \wedge P_2$

In 3D, Conformal Geometric Algebra (CGA) provides a great variety of basic geometric entities to compute with, namely points, spheres, planes, circles, lines, and point pairs. Table 15.2 shows the extension of the 2D geometric objects of the previous chapters to 3D objects. These entities have two algebraic representations: the IPNS (inner product null space) and the OPNS (outer product null space).[4] They are duals of each other (a superscript asterisk denotes the dualization operator). In Table 15.2, $\mathbf{x}$ and $\mathbf{n}$ are in bold type to indicate that they represent 3D entities obtained by linear combinations of the 3D basis vectors e_1, e_2, and e_3:

$$\mathbf{x} = x_1e_1 + x_2e_2 + x_3e_3. \tag{15.1}$$

The $\{S_i\}$ represent different spheres, and the $\{\pi_i\}$ represent different planes. In the OPNS representation, the outer product "$\wedge$" indicates the construction of a geometric object with the help of points $\{P_i\}$ that lie on it. A sphere, for instance, is defined by four points $(P_1 \wedge P_2 \wedge P_3 \wedge P_4)$ on this sphere. In the IPNS representation, the meaning of the outer product is an intersection of geometric entities. A circle, for instance, is defined by the intersection of two spheres $S_1 \wedge S_2$ (see Fig. 15.2). Just like with GAALOPScripts, with

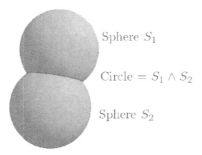

Sphere S_1

Circle $= S_1 \wedge S_2$

Sphere S_2

FIGURE 15.2 CLUCalc visualization with annotations: the intersection of two spheres results in a circle.

CLUScripts there is an almost one-to-one correspondence between formulas and code. The formulas

$$S_1 = P_1 - \frac{1}{2}r_1^2e_\infty,$$

$$S_2 = P_2 - \frac{1}{2}r_2^2e_\infty,$$

and

$$z = S_1 \wedge S_2$$

are coded in CLUCalc as follows:

```
S1 = P1 - 0.5*r1*r1*einf;
S2 = P2 - 0.5*r2*r2*einf;
z = S1^S2;
```

---

[4]Please refer to Sect. 5.2 for more details about these representations.

15.3 ANGLES AND DISTANCES IN 3D

The results for circles and lines in Sect. 7 hold for spheres and planes in Conformal Geometric Algebra. The equations indicated in Table 15.3 can be

TABLE 15.3 Geometric meaning of the inner product of (normalized) planes, spheres and points.

·	Plane	Sphere	Point
Plane	Angle between planes Eq. (7.9)	Euclidean distance from center, Eq. (7.13)	Euclidean distance Eq. (7.6)
Sphere	Euclidean distance from center, Eq. (7.13)	Distance measure Fig. 7.7	Distance measure Eq. (7.16)
Point	Euclidean distance Eq. (7.6)	Distance measure Eq. (7.16)	Distance Eq. (7.3)

transferred to 3D by extending the 2D coordinates to 3D coordinates. Please refer to Sect. 15.7 for a visibility application in 3D using distance computations based on the inner product of spheres.

15.4 3D TRANSFORMATIONS

Reflections as the basis of 3D transformations are handled in Conformal Geometric Algebra comparable to reflections in 2D according to Chapt. 8. The reflection of an object o at a plane P is expressed by

$$o_{reflecteded} = PoP. \tag{15.2}$$

Transformations such as rotations and translations are also represented by combinations of reflections. The operator

$$R = e^{-(\frac{\phi}{2})L} \tag{15.3}$$

describes a **rotor**. L is the rotation axis, represented by a normalized bivector, and ϕ is the rotation angle around this axis. R can also be written as

$$R = \cos\left(\frac{\phi}{2}\right) - L\sin\left(\frac{\phi}{2}\right). \tag{15.4}$$

The rotation of a geometric object o is performed with the help of the operation

$$o_{rotated} = Ro\tilde{R}.$$

There are strong relations between rotations in Conformal Geometric Algebra and quaternions and dual quaternions (see [26]).

In Conformal Geometric Algebra, a translation can be expressed in a multiplicative way with the help of a **translator** T defined by

$$T = e^{-\frac{1}{2}\mathbf{t}e_\infty}, \tag{15.5}$$

where $\mathbf{t}$ is a vector

$$\mathbf{t} = t_1 e_1 + t_2 e_2 + t_3 e_3.$$

Application of the Taylor series

$$T = e^{-\frac{1}{2}\mathbf{t}e_\infty} = 1 + \frac{-\frac{1}{2}\mathbf{t}e_\infty}{1!} + \frac{(-\frac{1}{2}\mathbf{t}e_\infty)^2}{2!} + \frac{(-\frac{1}{2}\mathbf{t}e_\infty)^3}{3!} + \cdots$$

and the property $(e_\infty)^2 = 0$ results in the translator

$$T = 1 - \frac{1}{2}\mathbf{t}e_\infty. \tag{15.6}$$

Example: Let us, for instance, translate the sphere

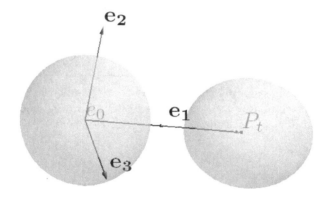

FIGURE 15.3 Translation of a sphere from the origin to the point P_t.

$$S = -e_\infty + e_0 \tag{15.7}$$

(see Fig. 15.3) in the x-direction by the translation vector

$$\mathbf{t} = 4e_1. \tag{15.8}$$

Note that this is a sphere with its center at the origin and with $r^2 = 2$.
    The translator in this example has the form

$$T = 1 - 2e_1 e_\infty, \tag{15.9}$$

and its reverse is

$$\tilde{T} = 1 + 2e_1 e_\infty. \tag{15.10}$$

The translated sphere can now be computed as the versor product

$$S_{translated} = T S \tilde{T} \tag{15.11}$$

$$= (1 - 2e_1 e_\infty)(-e_\infty + e_0)(1 + 2e_1 e_\infty)$$

$$= (1 - 2e_1 e_\infty)(-e_\infty - 2\underbrace{e_\infty e_1 e_\infty}_{0} + e_0 + 2e_0 e_1 e_\infty)$$

$$= (1 - 2e_1 e_\infty)(-e_\infty + e_0 - 2e_1 e_0 e_\infty)$$

$$= -e_\infty + e_0 - 2e_1 e_0 e_\infty + 2e_1 \underbrace{e_\infty e_\infty}_{0} - 2e_1 e_\infty e_0 + 4e_1 e_\infty e_1 e_0 e_\infty$$

$$= -e_\infty + e_0 - 2e_1 \underbrace{(e_0 e_\infty + e_\infty e_0)}_{-2} + 4e_1 e_\infty e_1 e_0 e_\infty$$

$$= 4e_1 - e_\infty + e_0 + 4 \underbrace{e_1 e_\infty e_1}_{-e_\infty} e_0 e_\infty$$

$$= 4e_1 - e_\infty + e_0 - 4e_\infty \underbrace{e_0 e_\infty}_{-e_\infty \wedge e_0 - 1}$$

$$= 4e_1 - e_\infty + e_0 - 4 \underbrace{e_\infty(-e_\infty \wedge e_0 - 1)}_{-2e_\infty},$$

resulting in

$$S_{translated} = 4e_1 + 7e_\infty + e_0. \tag{15.12}$$

This is a sphere with the same radius $r^2 = 2$, but with a translated center point

$$P_t = \mathbf{t} + \frac{1}{2}t^2 e_\infty + e_0 = 4e_1 + 8e_\infty + e_0. \tag{15.13}$$

Please notice that in 3D a rigid body motion is still more general in the sense that it consists of a rotation around an arbitrary line in space together with a translation in the direction of this line.

Please refer, for instance, to [26] and to the recent publication [2].

15.5 CLUCALC IMPLEMENTATION OF THE SNAKE ROBOT CONTROL

The snake control application of Chapt. 14 was tested with CLUCalc. The following piece of code contains the definitions of basic objects:

```
DefVarsN3();\\
// COMPUTATION OF COORDINATES IN CONFIGURATION SPACE\\
// POSITION OF ROBOT CORR. TO THE COORDINATES\\
// Initial position of points Q1, Q2, Q3 and line L1 \\
Q1=VecN3(0,0,0);
L1=VecN3(0,0,1);
T0=TranslatorN3(2,0,0);
Q2=T0*Q1*~T0;
Q3=T0*Q2*~T0;
Q4=T0*Q3*~T0;
// Computation of point pairs P1,P2 and P2
P1=Q1^Q2;
P2=Q2^Q3;
P3=Q3^Q4;
```

The initial position is thus recalculated with respect to the controlling parameters change.

```
\\ Coordinates x and y from configutation space
T=TranslatorN3(x,y,0);
\\ Axis of rotation in point (x,y)
LB=T*L1*~T;
\\ Rotor in space with respect to R
MB=TranslatorN3(LB)*RotorN3(0,0,1,d)*~TranslatorN3(LB);
\\ New position of point pair P1
:P1=MB*T*P1*~T*~MB;
\\ Projection to the first point of P1
T1=(-(sqrt(P1.P1)+P1)/(einf.P1));
\\ Rotor in first joint
L2=TranslatorN3(T1)*L1*~TranslatorN3(T1);
M1=TranslatorN3(L2)*RotorN3(0,0,1,a)*~TranslatorN3(L2);
\\ New position of point pair P2
:P2=M1*MB*T*P2*~T*~MB*~M1;
\\ Projection to the second point of P3
T2=((sqrt(P3.P3)+P3)/(einf.P3));
\\ Rotor in second joint
L3=TranslatorN3(T2)*L1*~TranslatorN3(T2);
M2=TranslatorN3(L3)*RotorN3(0,0,1,b)*~TranslatorN3(L3);
\\ New position of point pair P3
:P3=M2*M1*MB*T*P3*~T*~MB*~M1*~M2;
```

Fig. 15.4 demonstrates the evolution from 0 in the direction of the vector field g_1, i.e. when the controlling parameter t_1 is set to zero and t_2 is changed within the range $\langle 0, 2\pi \rangle$. The last figure shows the motion corresponding to the bracket $[g_1, g_2]$ which is realized by means of a periodic transformation of the generators g_1 and g_2:

$$v(t) = -\epsilon\omega \sin(\omega t)g_1 + \epsilon\omega \cos(\omega t)g_2,$$

where $\epsilon = 0.1$, $\omega = 4$ and $t \in \langle 0, 2\pi/\omega \rangle$.

FIGURE 15.4 $t_1 = 0, t_2 = t$ (visualized by CLUCalc).

FIGURE 15.5 Lie bracket $[g_1, g_2]$ (visualized by CLUCalc).

15.6 3D COMPUTATIONS WITH GAALOP

GAALOP can also be used for 3D computations based on Conformal Geometric Algebra. Please select "5d - conformal space" for "Algebra to use". The following GAALOPScript computes a line through two arbitrary points $p1$ and $p2$

Listing 15.1 *Line3D.clu*: Script for the computation of the line through two points $p1$ and $p2$.

```
1  // this is a GAALOPScript for 5d - conformal space
2
3  p1 = createPoint(px1,py1,pz1);
4  p2 = createPoint(px2,py2,pz2);
5
6  ?L = *(p1^p2^einf);
```

and results in the following C/C++ code

Listing 15.2 *Line3D.c*: Script for the computation of the line through two points $p1$ and $p2$.

```
1  L[6]  = pz1 - pz2; // e1 ^ e2
2  L[7]  = py2 - py1; // e1 ^ e3
3  L[8]  = py2 * pz1 - py1 * pz2; // e1 ^ einf
4  L[10] = px1 - px2; // e2 ^ e3
5  L[11] = px1 * pz2 - px2 * pz1; // e2 ^ einf
6  L[13] = px2 * py1 - px1 * py2; // e3 ^ einf
```

describing the simple arithmetic computations for 6 coefficients needed for the line multivector. If we extend this script in order to compute the line perpendicular to the x-y-plane at the 2D location (px1,py1) according to the following GAALOPScript,

Listing 15.3 *PointOfRotation3D.clu*: Script for the computation of the point of rotation at the 2D location (px1,py1).

```
1  // this is a GAALOPScript for 5d - conformal space
2
3  pz1=0;
4  px2=px1;
5  py2=py1;
6  pz2=1;
7
8  p1 = createPoint(px1,py1,pz1);
9  p2 = createPoint(px2,py2,pz2);
10
11 ?L = *(p1^p2^einf);
```

the resulting C code

Listing 15.4 *Line3D.c*: Script for the computation of the line through two points *p1* and *p2*.

```
1  L[6]  = -1.0; // e1 ^ e2
2  L[8]  = (-py1); // e1 ^ einf
3  L[11] = px1; // e2 ^ einf
```

corresponds exactly to the point of rotation of the snake robot application according to Eq. (14.5) when using the relevant 2D point (j,0).

15.7 VISIBILITY APPLICATION IN 3D

Here, we expand the visibility application of Chapt. 11 to 3D. The following GAALOPScript computes the inner product (as a measure of distance) of a bounding sphere with center (q1,q2,q3) and radius r2 to a number of spheres representing a view cone. It is modeled by an (observer) point at (p1x,p1y,p1z) and increasing spheres with center (p2x,p2y,p2z) and radius r1 at the end.

Listing 15.5 Computation of the inner product of a sphere and a cone modeled by spheres.

```
1
2  r=t*r1;
3  px = p1x + t*(p2x-p1x);
4  py = p1y + t*(p2y-p1y);
5  pz = p1z + t*(p2z-p1z);
6  P = createPoint(px,py,pz);
7  C1 = P - 0.5*r*r*einf;
```

```
 8
 9  Q = createPoint(q1,q2,q3);
10  C2 = Q - 0.5*r2*r2*einf;
11  ?Distance = 2*C1.C2;
```

It results in the following expression for the distance of the data sphere in dependence of the parameter t for the description of the spheres which are modeling the view cone:

Distance(t) = (((((((((r1 * r1 - p2z * p2z + 2.0 * p1z * p2z) - p2y * p2y + 2.0 * p1y * p2y) - p2x * p2x + 2.0 * p1x * p2x) - p1z * p1z - p1y * p1y - p1x * p1x) *** t * t** + (((2.0 * p2z - 2.0 * p1z) * q3 + (2.0 * p2y - 2.0 * p1y) * q2 + (2.0 * p2x - 2.0 * p1x) * q1) - 2.0 * p1z * p2z - 2.0 * p1y * p2y - 2.0 * p1x * p2x + 2.0 * p1z * p1z + 2.0 * p1y * p1y + 2.0 * p1x * p1x) *** t** + r2 * r2) - q3 * q3 + 2.0 * p1z * q3) - q2 * q2 + 2.0 * p1y * q2) - q1 * q1 + 2.0 * p1x * q1) - p1z * p1z - p1y * p1y - p1x * p1x; which is a polynomial in t (dependent on the maximum radius r1 of the cone, the radius r2 of the bounding sphere, the starting point p1x, p1y, p1z and the end point p2x, p2y, p2z of the cone).

Computing the first derivative results in

```
2*(r1^2-p2z^2+2.0*p1z*p2z-p2y^2+2.0*p1y*p2y-p2x^2+2.0*p1x*p2x
-p1z^2-p1y^2-p1x^2)*t+(2.0*p2z-2.0*p1z)*q3+(2.0*p2y-2.0*p1y)*q2
+(2.0*p2x-2.0*p1x)*q1-2.0*p1z*p2z-2.0*p1y*p2y-2.0*p1x*p2x+
2.0*p1z^2+2.0*p1y^2+2.0*p1x^2
```

The extremum is reached for $t = -p/q$ with

```
p=(p2z-p1z)*q3+(p2y-p1y)*q2+(p2x-p1x)*q1
-p1z*p2z-p1y*p2y-p1x*p2x+p1z^2+p1y^2+p1x^2
```

and

```
q=3*r2^2+r1^2-3*q3^2+(3*p2z+3*p1z)*q3-3*q2^2
+(3*p2y+3*p1y)*q2-3*q1^2+(3*p2x+3*p1x)*q1-p2z^2
-p1z*p2z-p2y^2-p1y*p2y-p2x^2-p1x*p2x-p1z^2-p1y^2-p1x^2
```

Note: if you are interested in some kind of average distance from the data sphere to the spheres of the view cone, you can use the integral between 0 and 1 resulting in

```
3*r2^2+r1^2-3*q2^2+(3*p2y+3*p1y)*q2-3*q1^2
+(3*p2x+3*p1x)*q1-p2y^2-p1y*p2y-p2x^2-p1x*p2x-p1y^2-p1x^2
```

15.8 CONCLUSION OF THE ENGINEERING PART

The 3D considerations of this chapter complete the engineering part of the book, hopefully inspiring many people to use Geometric Algebra for their applications. You can find an overview over Geometric Algebra applications, for instance, in [36]. Applications mainly from computer graphics, computer vision and robotics can be found in the books [26], [57], [8], [1] and [48]. There is a wide range of engineering applications able to benefit from Geometric Algebra. One idea for the future is to reformulate the algorithms of the well-known Graphics Gem book series. This could provide its own merit as "Geometric Algebra Gems" and promote Geometric Algebra over standard algorithms.

The following SECTION IV is added to give some considerations about using Geometric Algebra already at school and about *Space-Time Algebra* in honor of the work of David Hestenes and especially the 50th anniversary of his book about this algebra.

IV

Geometric Algebra at School

CHAPTER 16

Geometric Algebra for Mathematical Education

CONTENTS

SECTION IV gives some considerations about using Geometric Algebra already at school and about *Space-Time Algebra* in honor of the work of David Hestenes and especially to the 50th anniversary of his book about this algebra. The professional life of David Hestenes is very much the professional life of an educator. He has held the Chair of Physics Education at Arizona State University. Thus he invented, improved and implemented Geometric Algebra first as a tool in physics education [18], [19] and undoubtedly in mathematics education [21]. Thus David Hestenes' lines of thought always started with didactical questions and only then reached the field of Geometric Algebra as a proper improvement of our present-day mathematical views and a proper answer to our present-day mathematical problems.

Since it focuses on the most basic geometric objects, lines and circles, Compass Ruler Algebra hopefully can help to introduce Geometric Algebra already in school. While there are already very good Dynamic Geometry Systems (DGS) such as Geogebra and Cinderella [1] this chapter focuses on the question of whether GAALOP can be the base for an appropriate tool for

---

[1] The work of Eckhard Hitzer et al. [37] dealing with Geometric Algebra using Cinderella is one of the inspirations of this book.

mathematical education based on Geometric Algebra. It mainly reviews some thoughts from the paper [32].

16.1 BASIC DGS FUNCTIONALITY BASED ON GAALOP

Compass Ruler Algebra provides a strong relation between geometry and algebra. In this section we review some basic DGS functionality based on GAALOP and this algebra.

Geometric Algebra is a very general mathematical system providing simultaneously a geometrification of algebra, and also an algebraification of geometry. Compass Ruler Algebra as presented in this book is very well suited to compute similar to working with compass and ruler[2]. Geometric objects such as circles and lines as well as geometric operations with them can be handled very easily inside of the algebra. A circle, for instance, can be described based on the outer product of three points of the circle. You are able to directly compute with infinity, for instance, when expressing the center point of a circle as the inversion of infinity in the circle [3]. Taken together this shows how the objects of Compass Ruler Algebra can be described in an algebraic yet syntactic fashion so that a close link between algebra and geometry is established.

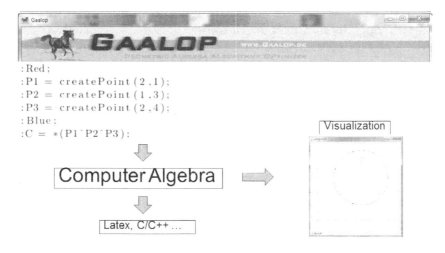

FIGURE 16.1 GAALOP symbolically computes Geometric Algebra expressions and generates geometry (visualizations) or simplified formulae (in LaTeX, C/C++ ... format).

---

[2]See Chapt. 3 and Chapt. 5
[3]See Sect. 8.8

GAALOP as presented in this book is an easy to handle tool in order to compute and visualize with Compass Ruler Algebra. While computer algebra functionality is responsible for the symbolic computations, its visualizing component offers basic DGS (Dynamic Geometry System) functionality. According to Fig. 16.1, we currently use GAALOP for visualizations and for the generation of formulae expressed in LaTeX or C++ format. Fig. 16.1 also shows how easy it is to compute and visualize with GAALOP. First of all, three points with the 2D coordinates $(2,1), (1,3)$ and $(2,4)$ are transformed into 4D coordinates of the Compass Ruler Algebra and visualized in red (all variables with a leading colon are visualized with the currently set color). Then the circle C is computed based on the outer product of these three points, transformed into the standard representation via the dualization operator and visualized in blue. Note: this way the circumcircle of a triangle can be computed very easy.

16.2 GEOMETRIC CONSTRUCTIONS BASED ON COMPASS RULER ALGEBRA

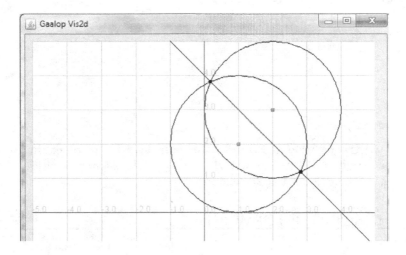

FIGURE 16.2 Visualization of the perpendicular bisector between the two red points.

The example of Sect. 3.2.5 shows how, in Compass Ruler Algebra, we are able to compute comparable to working with compass and ruler. In order to construct the perpendicular bisector of a section of a line with compass and ruler, we draw two circles with the center at the boundary points and connect the two intersection points according to Fig. 16.2. This can be expressed with the help of the GAALOPScript 16.1.

Listing 16.1 *PerpendicularBisector.clu*: Script for the computation of the perpendicular bisctor.

```
1  x1 = 1;
2  y1 = 2;
3  x2 = 2;
4  y2 = 3;
5  r = 2;
6
7  P1 = createPoint(x1,y1);
8  P2 = createPoint(x2,y2);
9  // intersect two circles with center points P1 and P2
10 // with the same, but arbitrary radius
11 S1 = P1 - 0.5*r*r*einf;
12 S2 = P2 - 0.5*r*r*einf;
13 PPdual = *(S1^S2);
14 // the line through the two points
15 // of the resulting point pair
16 Bisector = *(PPdual^einf);
17
18 :Red;
19 :P1;
20 :P2;
21 :Black;
22 :S1;
23 :S2;
24 :Bisector;
```

For the visualization of Fig. 16.2 the variables have to be equipped first with concrete input values; as well the colors for the geometric objects to be drawn have to be defined at the end.

16.3   DERIVING OF FORMULAE

GAALOP can also be used to derive formulae of geometric relationships. In case we use variable names in Sect. 3.2.5 instead of concrete values for the point coordinates, GAALOP computes symbolically and the resulting formulae are presented in output formats such as C code or LaTeX code. Listing 16.2 makes exactly that.

Listing 16.2 *PerpendicularBisectorCode.clu*: Script for the computation of the perpendicular bisctor.

```
1  P1 = createPoint(x1,y1);
2  P2 = createPoint(x2,y2);
3  // intersect two circles with center points P1 and P2
4  // with the same, but arbitrary radius
```

```
 5  S1 = P1 - 0.5*r*r*einf;
 6  S2 = P2 - 0.5*r*r*einf;
 7  PPdual = *(S1^S2);
 8  // the line through the two points
 9  // of the resulting point pair
10  ?Bisector = *(PPdual^einf);
```

Around the two points with the symbolic 2D coordinates (x1,y1) and (x2,y2) two circles are drawn with radius r and the perpendicular bisector is computed based on the line through the two intersecting points of the circles. The result of Listing 16.2 is the multivector Bisector expressed as follows (please refer to Table 2.2 for the indices of the multivector).

$$Bisector_1 = x2 - x1$$
$$Bisector_2 = y2 - y1$$
$$Bisector_3 = \frac{y2 * y2}{2} - \frac{y1 * y1}{2} + \frac{x2 * x2}{2} - \frac{x1 * x1}{2}$$

What we immediately see is that the resulting multivector consists only of e_1, e_2, e_∞ coefficients and no e_0 component, meaning the result is a line. Its normal vector is the difference of the 2D-vectors (x1,x2) and (y1,y2). This means that the line is perpendicular to the direction vector of (x1,x2) and (y1,y2).

We may assume now, that the multivector Bisector can also be expressed simply as the difference of the two points, especially since in our example of Sect. 3.2.6 the difference of two points computes the line in the middle of two points. Is that true in arbitrary cases? The following GAALOPScript

Listing 16.3 *DifferencePointPointCode.clu*: Script for the computation of the difference of 2 points.

```
1  P1 = createPoint(x1,y1);
2  P2 = createPoint(x2,y2);
3  ?Diff = P2-P1;
```

computes the difference of the two points P2 and P1 and results in

$$Diff_1 = x2 - x1$$
$$Diff_2 = y2 - y1$$
$$Diff_3 = \frac{y2 * y2}{2} - \frac{y1 * y1}{2} + \frac{x2 * x2}{2} - \frac{x1 * x1}{2}$$

which is exactly the same as the result of the multivector Bisector. In this way we derived the formula

$$B = P_2 - P_1 \qquad\qquad (16.1)$$

for the perpendicular bisector B of the line segment from the point P_1 to the point P_2.

16.4 PROVING GEOMETRIC RELATIONSHIPS

Here we show how easy it is to prove geometric relationships with GAALOP. According to Fig. 16.3 we prove based on GAALOP that the bisectors of the line segments from p0 to p1 and from p0 to p2 intersect in the center point of the circle going through all the three points p0, p1 and p2.

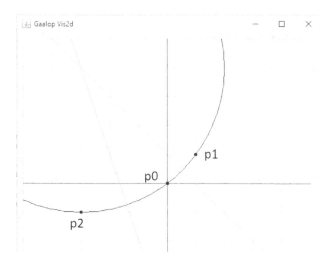

FIGURE 16.3 The bisectors of the line segments from p0 to p1 and from p0 to p2 intersect in the center point of the circle going through all the three points p0, p1, p2.

We use the following GAALOPScript based on the scripts of Chapt. 10 and Chapt. 12

Listing 16.4 *Proof.clu.*

```
1  p0 = e0;   // p0 at the origin
2  p1 = createPoint(px1,py1);
3  p2 = createPoint(px2,py2);
4  l1 = p1-p0;   // line in the middle of p1 and p0
5  l2 = p2-p0;   // line in the middle of p2 and p0
6  PpOPNS = l1^l2;
7  Pp=*PpOPNS;
8  IP = Pp.e0;
```

```
 9
10  ?IPp0 = IP.p0;  // p0 on the circle IP?
11  ?IPp1 = IP.p1;  // p1 on the circle IP?
12  ?IPp2 = IP.p2;  // p0 on the circle IP?
```

in order to prove that.

In our example of Chapt. 10 we show that IP is not only the center point of the circle going through all the three points p0, p1, p2, but already the complete circle. If we would like to prove that for arbitrary points p1 and p2 (p0 is always at the origin), we have to show that p0, p1 and p2 are in the inner product null space (IPNS) of the circle. This can be done based on the inner products of IP with all three points (lines 10 - 13). The result of these operations is zero proving that all three points are lying on the circle IP.

16.5   OUTLOOK

The previous sections showed that GAALOP is able to provide basic DGS functionality based on Compass Ruler Algebra. This algebra is able to handle basic geometric objects and operations while always combining geometry and algebra in a very consistent way.

Nevertheless, there are open items. A core question of Dynamic Geometry Systems, as stated in the PhD thesis of Kortenkamp [49] is: How should a construction behave under movements? The most natural thing would be a continuous movement of dependent elements: When a free element is moved only a little bit, then the dependent elements will move only a little bit, too. We do not want elements to jump around wildly. However, it has been proven by Kortenkamp that a moving strategy can either be continuous or deterministic. So the question arises whether a geometric algebra produces continuous or deterministic behavior. However, as e.g. intersections of circles produce point pairs, there is no obvious strategy for how to select a point from a point pair - and exactly this decision is the one that decides between continuous or deterministic behavior. One important situation is two moving circles. In Compass Ruler Algebra, their intersections are not two different points but a point pair as only one algebraic expression (see Section 5.11). The intersection of two circles can be investigated based on the example of computing the perpendicular bisector (see Figure 3.9). This example raises questions such as

> what happens, if the radius is smaller than half the distance of the two points?

> what about an imaginary radius?

> what happens with the point pair, if the two circles touch each other?

> what happens with the point pair, if the two circles overlap each other?

With Geometric Algebra, there is a good chance of answering these questions in a consistent way.

CHAPTER 17

Space-Time Algebra in School and Application

CONTENTS

In honor of the 50th anniversary of the book *Space-Time Algebra* of David Hestenes, this chapter gives an introduction on how to compute with Space-Time Algebra as well as an application based on GAALOP. Interestingly, Space-Time Algebra and Compass Ruler Algebra as mainly used in this book have a similar algebraic structure.

17.1   THE ALGEBRAIC STRUCTURE OF SPACE-TIME ALGEBRA

The basis of Space-Time Algebra consists of one time-like vector (squaring to +1) and three space-like vectors (squaring to -1). These basis vectors can be identified with (or at least represented by) Dirac Gamma Matrices. Therefore the Greek letter γ is usually used to express space-time basis vectors.

As time-like and space-like basis vectors have different signatures, this mathematical design describes a hyperbolic, Pseudo-Euclidean geometry in

TABLE 17.1   The 4 basis vectors of the Space-Time Algebra.

	basis vector	Signature	GAALOP
γ_t (time-like)	e_1	+1	gt
γ_x (space-like)	e_2	-1	gx
γ_y (space-like)	e_3	-1	gy
γ_z (space-like)	e_4	-1	gz

TABLE 17.2 The 16 basis blades of the Space-Time Algebra.

Index	Pauli/Dirac Matrices	Blade	Dimension	Square
0	1	1	0 scalar	+1
1	γ_t	e_1	1 vector	+1
2	γ_x	e_2	1 vector	-1
3	γ_y	e_3	1 vector	-1
4	γ_z	e_4	1 vector	-1
5	$\gamma_t\gamma_x = -\sigma_x$	$e_1 \wedge e_2$	2 bivector	+1
6	$\gamma_t\gamma_y = -\sigma_y$	$e_1 \wedge e_3$	2 bivector	+1
7	$\gamma_t\gamma_z = -\sigma_z$	$e_1 \wedge e_4$	2 bivector	+1
8	$\gamma_x\gamma_y = -\sigma_x\sigma_y$	$e_2 \wedge e_3$	2 bivector	-1
9	$\gamma_x\gamma_z = -\sigma_x\sigma_z$	$e_2 \wedge e_4$	2 bivector	-1
10	$\gamma_y\gamma_z = -\sigma_y\sigma_z$	$e_3 \wedge e_4$	2 bivector	-1
11	$\gamma_t\gamma_x\gamma_z$	$e_1 \wedge e_2 \wedge e_3$	3 trivector	-1
12	$\gamma_t\gamma_x\gamma_z$	$e_1 \wedge e_2 \wedge e_4$	3 trivector	-1
13	$\gamma_t\gamma_y\gamma_z$	$e_1 \wedge e_3 \wedge e_4$	3 trivector	-1
14	$\gamma_x\gamma_y\gamma_z$	$e_2 \wedge e_3 \wedge e_4$	3 trivector	+1
15	$\gamma_t\gamma_x\gamma_y\gamma_z = \sigma_x\sigma_y\sigma_z$	$e_1 \wedge e_2 \wedge e_3 \wedge e_4$	4 quadvector	-1

contrast to the Euclidean geometry of ordinary three-dimensional space. Therefore vectors, which square to 0, exist. They are called light-like vectors. The relation to basis vectors of 3-dimensional Geometric Algebra which can be identified with (or at least represented by) Pauli Matrices $\sigma_x, \sigma_y, \sigma_z$ (squaring to +1) is given by the remarkable formula

$$\sigma_k = \gamma_k\gamma_t \tag{17.1}$$

[19] (p. 695, eq. 43) which is a highlight [64] (p. 1292) of the Geometric Algebra picture of special relativity.

17.2 SPACE-TIME ALGEBRA AT SCHOOL

Inside the Geometric Algebra community there is a vivid discussion about how to implement Geometric Algebra and Space-Time Algebra at school and high school [56]. Martin E. Horn recently applied Space-Time Algebra at high school [43, 40, 41] and introductory courses at universities of applied sciences [42] for teaching special relativity with topics such as

Time dilation

Twin paradox

Length contraction

Lorentz transformation

We will use GAALOP together with his basic examples for the description of Space-Time Algebra[1]. GAALOP can be used for algebras different from the Compass Ruler Algebra as mainly used in this book. One predefined option is the Space-Time Algebra of special relativity. According to Fig. 17.1 the only thing we have to do is to change the algebra to be used to **st4d-space-time**.

FIGURE 17.1   GAALOP configured for Space-Time Algebra.

Computing the square of space-time vector r with the following GAALOP-Script

Listing 17.1   *STASquare.clu*: Computation of the square of a space-time vector.

```
1  gt = e1;
2  gx = e2;
3  gy = e3;
4  gz = e4;
5  r = r1*gt + r2*gx + r3*gy + r4*gz;
6  ?IP=r.r;
```

results in the following formula

$$IP_0 = r1 * r1 - r2 * r2 - r3 * r3 - r4 * r4$$

representing the square of the length of vector r which will be positive if vector r is a time-like vector, negative if vector r is a space-like vector, and 0 if vector r is a light-like vector.

Computing the inner and outer products of a space-time vector r (which describes an event at position r) and a unit vector n (which describes the time axis of the coordinate system of an observer) with the following GAALOP-Script

---

[1]See [44] for an example to use GAALOP at school and high school.

Listing 17.2 *STAProducts.clu*: Computation of the inner and outer products of a space-time vectors.

```
1  gt  =  e1;
2  gx  =  e2;
3  gy  =  e3;
4  gz  =  e4;
5  r  =  r1*gt  +  r2*gx  +  r3*gy  +  r4*gz;
6  n  =  n1*gt  +  n2*gx  +  n3*gy  +  n4*gz;
7  ?IP=r.n;
8  ?OP=r^n;
```

results in the following formulae (see Fig. 17.2 for the indices of multivector OP)

$$IP_0 = r1 * n1 - r2 * n2 - r3 * n3 - r4 * n4$$
$$OP_5 = r1 * n2 - r2 * n1$$
$$OP_6 = r1 * n3 - r3 * n1$$
$$OP_7 = r1 * n4 - r4 * n1$$
$$OP_8 = r2 * n3 - r3 * n2$$
$$OP_9 = r2 * n4 - r4 * n2$$
$$OP_{10} = r3 * n4 - r4 * n3$$

representing the n-split of space-time. The inner product $r \cdot n$ assigns a unique time to every event r [2, p. 395, eq. 38], measured by the observer. And the outer product $r \wedge n$ assigns a unique position to every event r [2, p. 395, eq. 39], measured by the observer.

The following GAALOPScript with three space-time vectors a, b, c according to Fig. 17.2

Listing 17.3 *STAOrthogonality.clu*: Computation of the inner product of two space-time vectors.

```
1  gt  =  e1;
2  gx  =  e2;
3  a  =  a0*gt  +  a1*gx;
4  b  =  a1*gt  +  a0*gx;
5  c  =  -a1*gt  +  a0*gx;
6  ?ac=a.c;
7  ?ab=a.b;
```

shows a surprising result:

$$ac_0 = -2 * a0 * a1$$
$$ab_0 = 0$$

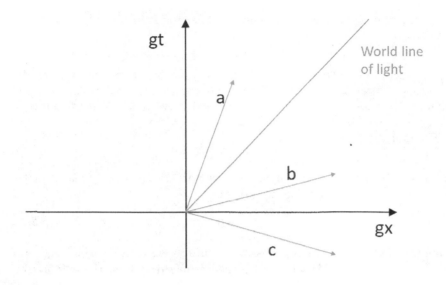

FIGURE 17.2   Orthogonality of space-time vectors.

Not the two vectors a and c are orthogonal but the vectors a and b, and this is a simple geometric consequence [51] (p. 43, fig. 25) of the hyperbolic geometry of Special Relativity and our problematic attempts to visualize this Pseudo-Euclidean situation on a Euclidean piece of paper.

17.3   A FARADAY EXAMPLE FOR MATHEMATICA'S OpenCLLink

This section demonstrates how to use the GAALOP precompiler for applications such as the computation of the electromagnetic field generated by a moving charged point particle capturing relativistic effects (see [3]).

In order to capture relativistic effects, four-dimensional space-time is employed. An analytic treatment of this example is explained in section 7.3 of the book [7][2].

In [3], the following sign convention[3] is chosen for Space-Time Algebra:

$$e_4^2 = +1$$

and

$$e_1^2 = e_2^2 = e_3^2 = -1.$$

To translate three-dimensional vector, e.g., electric field, to its space-time form we proceed as follows:

$$\vec{E} = E^x \sigma_x + E^y \sigma_y + E^z \sigma_z = E^x e_1 e_4 + E^y e_2 e_4 + E^z e_3 e_4.$$

---

[2]See also [60] for physics applications.

[3]Note that in physics the signatures for the basis vectors + + + - and - - - + are equivalent for all practical purposes.

The extraction of, e.g., E^y, from Faraday bivector $F = \vec{E} + I\vec{B}$ is obtained by calculating

$$E^y = (e_2 e_4) \cdot F.$$

The vector $r_0 = [x_0, y_0, z_0, t_0]$ is the position of our moving particle in space-time, where x,y,z denote the spatial coordinates and t the time coordinate. Its trajectory $r_0(\tau) = [x_0(\tau), y_0(\tau), z_0(\tau), t_0(\tau)]$ is parameterized through the so called proper time τ.

The null vector

$$X = x - x_0(\tau) \tag{17.2}$$

connects an arbitrary point x in space-time with the trajectory $x_0(\tau)$. Since this vector is null by definition, we use

$$X^2 = 0 \tag{17.3}$$

to obtain $\tau(x)$. From the two possible solutions we only consider the one which is in the past. This solution is called retarded and has the property of $t_0(\tau) < t$.

With the knowledge of $\tau(x)$, one may calculate the electromagnetic field bivector

$$F(\tau) = \frac{X \wedge v + \frac{1}{2} X \dot{v} \wedge vX}{(X \cdot v)^3} \tag{17.4}$$

where $X = X(\tau), v = v(\tau), \dot{v} = \dot{v}(\tau)$ are all functions of τ, with

$$v = v(\tau) = \frac{dx_0}{d\tau} \tag{17.5}$$

being the velocity of the particle and

$$\dot{v} = \dot{v}(\tau) = \frac{d^2 x_0}{d\tau^2} \tag{17.6}$$

being its acceleration in space-time.

Having F it is easy to implement the following OpenCL-kernel:

```
__kernel void faraday_kernel(
          __global float* toMathematica,
          __global float* fromMathematica,
          const int length)
{
          const size_t index = get_global_id(0);
          if(index >= length)
                    return;

#pragma gpc begin
          float Xx = fromMathematica[index*12 + 0];
          float Xy = fromMathematica[index*12 + 1];
```

```
13          float Xz = fromMathematica[index*12 +  2];
14          float Xt = fromMathematica[index*12 +  3];
15          float Vx = fromMathematica[index*12 +  4];
16          float Vy = fromMathematica[index*12 +  5];
17          float Vz = fromMathematica[index*12 +  6];
18          float Vt = fromMathematica[index*12 +  7];
19          float Vdotx = fromMathematica[index*12 +  8];
20          float Vdoty = fromMathematica[index*12 +  9];
21          float Vdotz = fromMathematica[index*12 + 10];
22          float Vdott = fromMathematica[index*12 + 11];
23   #pragma clucalc begin
24          X = Xx*e1+Xy*e2+Xz*e3+Xt*e4;
25          V = Vx*e1+Vy*e2+Vz*e3+Vt*e4;
26          Vdot = Vdotx*e1+Vdoty*e2+Vdotz*e3+Vdott*e4;
27
28          dot = X.V;
29          ?F = (X^V+0.5f*X*Vdot^V*X)/(dot*dot*dot);
30   #pragma clucalc end
31          (toMathematica+index) = mv_to_array(F,
32      1,e1,e2,e3,e4,e1^e2,e1^e3,e1^e4,e2^e3,e2^e4,e3^e4,
33      e1^(e2^e3),e1^(e2^e4),e1^(e3^e4),e2^(e3^e4),
34      e1^(e2^(e3^e4)));
35   #pragma gpc end
36   }
```

This reads all values from an array called fromMathematica. It then calculates F and saves it to a second array called toMathematica.

The code may be compiled and loaded using the following Mathematica-commands:

```
 1   (*import OpenCLLink*)
 2   Needs["OpenCLLink'"]
 3
 4   (*system command for precompilation with GAALOP GPC*)
 5   command =
 6   "java -jar starter-1.0.0.jar -algebraName st4d"
 7   "-m usr/bin/maxima -optimizer de.GAALOP.tba.Plugin"
 8   "-generator de.GAALOP.compressed.Plugin"
 9   "-o \"out.cl\" -i \"in.clg\""
10
11   (*export code to file, precompile and import*)
12   Export["in.clg", code]
13   Run[command]
14   code = Import["out.cl"];
15
16
```

```
17  (*compile and link the OpenCL-kernel
18     to the function faraday*)
19  faraday = OpenCLFunctionLoad[code, "faraday_kernel",
20          {{_Real} , {_Real} , {_Integer} , {16} ,
21          "ShellOutputFunction" -> Print]
```

As you can see, with the statement -algebraName st4d, the code is compiled using the four dimensional Space-Time geometric algebra. This example shows that GAALOP can handle geometric algebras of arbitrary dimension and signature. This was a contribution of [66]: GAALOP needs only two definition files for using a Geometric Algebra. The well-known Conformal Geometric Algebra, the Euclidean and Projective Geometric Algebra and some others are already included in GAALOP besides of the four dimensional Space-Time Geometric Algebra.

They are specified respectively by putting

-algebra name 5d,

-algebraName 3d

and

-algebraName 4d

in the GAALOP command. Please refer to [66] and the GAALOP-manual at [29] for more information on this topic.

The OpenCLLink-function **faraday** may be used to compute F over a range and to plot it subsequently.

```
1   (*constants*)
2   alpha = 1;
3   Omega = Pi;
4   startTime = -20;
5
6   (*define the space-time range to be evaluated*)
7   t0[tau_] = tau*Cosh[alpha];
8   x0[tau_] = (1/Omega)*Cos[Omega*tau]*Sinh[alpha];
9   y0[tau_] = (1/Omega)*Sin[Omega*tau]*Sinh[alpha];
10  z0[tau_] = 0;
11
12  (*compute tau table*)
13  range = 10;
14  tab = ParallelTable[temp[{t, x, y, z},
15     tau /.
16     FindRoot[(t - t0[tau])^2 - (x - x0[tau])^2 -
17        (y - y0[tau])^2 - (z - z0[tau])^2,
18        {tau, startTime}]],
19        {t, -range, range, .5}, {x, -range, range, .5},
20        {y, -range, range, .5}, {z, -range, range, 2}];
21        // AbsoluteTiming
22  tab = Flatten[tab];
```

```
23  tab = tab /. temp -> List
24  tau = Interpolation[tab]
25
26  (*define the input arrays*)
27  r = {t, x, y, z} // RotateLeft[#, 1] &
28  r0 = {taus*Cosh[alpha],
29      (1/Omega)*Cos[Omega*taus]*Sinh[alpha],
30      (1/Omega)*Sin[Omega*taus]*Sinh[alpha], 0}
31      // RotateLeft[#, 1] &
32  v = D[r0, taus]
33  vdot = D[v, taus]
34  X[t_, x_, y_, z_] = N[r-r0]/.{taus->tau[t,x,y,z]}
35  V[t_, x_, y_, z_] = N[v] /. {taus -> tau[t, x, y, z]}
36  Vdot[t_, x_, y_, z_] = N[vdot]/.{taus->tau[t,x,y,z]}
37
38  (*run faraday OpenCL-kernel and save results to F*)
39  F[t_, x_, y_, z_] =
40   faraday[Sequence @@
41      Join[X[t, x, y, z], V[t,x,y,z], Vdot[t,x,y,z]]]
42
43  (*extract electric field E from F*)
44  ExEy[x_, y_] = (F[3.1, x, y, 0.]);
45
46  (*plot E*)
47  Block[{t = 0, z = 0},
48   DensityPlot[Norm[ExEy[x,y]],{x,-10,10},{y,-10,10},
49    PlotPoints -> 150, ImageSize -> 500]]
```

The OpenCLLink-kernel produces the data for Figure 17.3 showing the magnitude of the electric field $|\vec{E}| = \sqrt{E_x^2 + E_y^2}$ extracted from F.

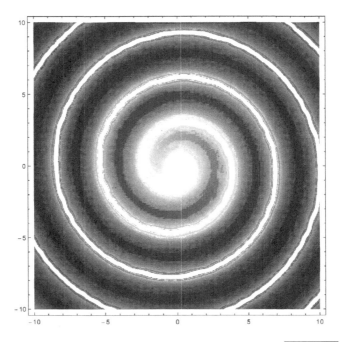

FIGURE 17.3 Magnitude of the electric field $|\vec{E}| = \sqrt{E_x^2 + E_y^2}$ of a rotating charge at relativistic velocity. Note that $E_z = 0$ due to the symmetry of the field.

Bibliography

[1] Eduardo Bayro-Corrochano. *Geometric Computing for Wavelet Transforms, Robot Vision, Learning, Control and Action.* Springer, 2010.

[2] Mauricio Belon and Dietmar Hildenbrand. Practical geometric modeling using geometric algebra motors. *Advances in Applied Clifford Algebras Journal*, 2017.

[3] Patrick Charrier, Mariusz Klimek, Christian Steinmetz, and Dietmar Hildenbrand. Geometric algebra enhanced precompiler for c++, opencl and mathematicas opencllink. *Advances in Applied Clifford Algebras Journal*, 2014.

[4] G. S. Chirikjian and J. W. Burdick. An obstacle avoidance algorithm for hyper-redundant manipulators. In *Proceedings 1990 IEEE International Conference on Robotics and Automation.*

[5] William Kingdon Clifford. *Applications of Grassmann's Extensive Algebra*, volume 1 of *American Journal of Mathematics*, pages 350–358. The Johns Hopkins University Press, 1878.

[6] Pablo Colapinto. The versor home page. Available at http://versor.mat.ucsb.edu/, 2015.

[7] Chris Doran and Anthony Lasenby. *Geometric Algebra for Physicists.* Cambridge University Press, 2003.

[8] Leo Dorst, Daniel Fontijne, and Stephen Mann. *Geometric Algebra for Computer Science, An Object-Oriented Approach to Geometry.* Morgan Kaufmann, 2007.

[9] K.J. Dowling. *Limbless Locomotion: Learning to Crawl with a Snake Robot.* PhD thesis, Carnegie Melon University, Pittsburgh, USA, 1997.

[10] Ahmad Hosney Awad Eid. *Optimized Automatic Code Generation for Geometric Algebra Based Algorithms with Ray Tracing Application.* PhD thesis, Suez Canal University, Port Said, 2010.

[11] Daniel Fontijne, Tim Bouma, and Leo Dorst. Gaigen 2: A geometric algebra implementation generator. Available at http://staff.science.uva.nl/~fontijne/gaigen2.html, 2007.

[12] Daniel Fontijne and Leo Dorst. Performance and elegance of 5 models of geometry in a ray tracing application. *Software and other downloads available at http://www.science.uva.nl/~fontijne/raytracer*, 2002.

[13] L. Gonzalez-Jimenez, O. Carbajal-Espinosa, A. Loukianov, and E. Bayro-Corrochano. Robust pose control of robot manipulators using conformal geometric algebra. *Algebra. Adv. App. Clifford Algebr.*, 24(2):533–552, 2014.

[14] Hermann Grassmann. *Die Ausdehnungslehre. Vollstaendig und in strenger Form begruendet.* Verlag von Th. Chr. Fr. Enslin, Berlin, 1862.

[15] David Hestenes. *Space-Time Algebra.* Gordon and Breach, New York, 1966.

[16] David Hestenes. *New Foundations for Classical Mechanics.* Springer, 1999.

[17] David Hestenes. Old wine in new bottles: A new algebraic framework for computational geometry. In Eduardo Bayro-Corrochano and Garret Sobczyk, editors, *Geometric Algebra with Applications in Science and Engineering.* Birkhäuser, 2001.

[18] David Hestenes. Oersted medal lecture 2002: Reforming the mathematical language of physics. *American Journal of Physics*, 71:104–121, 2003.

[19] David Hestenes. Spacetime physics with geometric algebra. *American Journal of Physics*, 71:691–714, 2003.

[20] David Hestenes. Grassmann's legacy. In H-J. Petsche, A. Lewis, J. Liesen, and S. Russ, editors, *From Past to Future: Grassmann's Work in Context.* Birkhäuser, 2011.

[21] David Hestenes. Modeling theory for math and science education. In Richard Lesh, Peter Galbraith, Christopher Haines, and Andrew Hurford (Eds.), editors, *Modeling Students Mathematical Modeling Competencies. Proceedings of the International Conferences on the Teaching of Mathematical Modeling and Applications ICTMA 13*, volume 1, pages 13–41. Springer Science + Business Media, Dordrecht, 2013.

[22] David Hestenes. *Space-Time Algebra (2nd edn.).* Birkhaeuser, New York, 2015.

[23] David Hestenes. The genesis of geometric algebra - a personal retrospective. In *Advances in Applied Clifford Algebras Journal*, open access at Springerlink.com, 2016.

[24] David Hestenes and Garret Sobczyk. *Clifford Algebra to Geometric Calculus: A Unified Language for Mathematics and Physics.* Springer, 1987.

[25] Dietmar Hildenbrand. *Geometric Computing in Computer Graphics and Robotics using Conformal Geometric Algebra.* PhD thesis, TU Darmstadt, 2006. Darmstadt University of Technology.

[26] Dietmar Hildenbrand. *Foundations of Geometric Algebra Computing.* Springer, 2013.

[27] Dietmar Hildenbrand, Werner Benger, and Yu Zhaoyuan. Practical geometric modeling using geometric algebra motors. *Advances in Applied Clifford Algebras Journal,* 2017.

[28] Dietmar Hildenbrand and Patrick Charrier. Conformal geometric objects with focus on oriented points. In *ICCA9, 7th International Conference on Clifford Algebras and Their Applications,* 2011.

[29] Dietmar Hildenbrand, Patrick Charrier, Christian Steinmetz, and Joachim Pitt. GAALOP home page. Available at http://www.gaalop. de, 2017.

[30] Dietmar Hildenbrand, Daniel Fontijne, Yusheng Wang, Marc Alexa, and Leo Dorst. Competitive runtime performance for inverse kinematics algorithms using conformal geometric algebra. In *Eurographics Conference Vienna,* 2006.

[31] Dietmar Hildenbrand, Silvia Franchini, A. Gentile, G. Vassallo, and S. Vitabile. Gappco: an easy to configure geometric algebra coprocessor based on gapp programs. *Advances in Applied Clifford Algebras Journal,* 2017.

[32] Dietmar Hildenbrand and Reinhard Oldenburg. Geometric algebra: a foundation for the combination of dynamic geometry systems with computer algebra systems? *The Electronic Journal of Mathematics and Technology,* 2015.

[33] S. Hirose. *Biologically Inspired Robots (Snake-like Locomotor and Manipulator).* Oxford University Press, 1993.

[34] Eckhard Hitzer. Tutorial on reflections in geometric algebra. In Kanta Tachibana, editor, *Lecture Notes of the International Workshop for Computational Science with Geometric Algebra,* 2007.

[35] Eckhard Hitzer. Angles between subspaces. In Vaclav Skala, editor, *Workshop Proceedings: Computer Graphics, Computer Vision and Mathematics 2010, Brno University of Technology,* 2010.

[36] Eckhard Hitzer, Tohru Nitta, and Yasuaki Kuroe. Applications of Clifford's geometric algebra. *Advances in Applied Clifford Algebras,* 23(2):377–404, 2013.

[37] Eckhard Hitzer and Luca Redaelli. Geometric algebra illustrated by cinderella. *Advances in Applied Clifford Algebras*, 2003.

[38] Eckhard Hitzer and Steve Sangwine. Multivector and multivector matrix inverses in real Clifford algebras. Technical report, University of Essex, 2016.

[39] Eckhard M. S. Hitzer. Euclidean geometric objects in the Clifford geometric algebra of Origin, 3-Space, Infinity. *Bulletin of the Belgian Mathematical Society - Simon Stevin*, 11(5):653–662, 2004.

[40] Martin E. Horn. Die spezielle Relativitaets-theorie im Kontext der Raumzeit-algebra. In *Didaktik der Physik, Beitraege zur Fruehjahrstagung in Bochum, Tagungs-CD des Fachverbands Didaktik der Physik der Deutschen Physikalischen Gesellschaft*, 2009.

[41] Martin E. Horn. Die Raumzeit-algebra im Abitur. In *PhyDid B Didaktik der Physik, Beitraege zur DPG-Fruehjahrstagung des Fachverbands Didaktik der Physik in Hannover*, 2010.

[42] Martin E. Horn. Die spezielle Relativitaetstheorie in der Mathematikerausbildung, Pauli-algebra und Dirac-algebra. oh-folien zur einfuehrung in die spezielle Relativittstheorie im rahmen eines kurses zur physik fuer mathematiker. In *PhyDid B Didaktik der Physik, Beitraege zur DPG-Fruehjahrstagung des Fachverbands Didaktik der Physik in Hannover*, 2010.

[43] Martin E. Horn. Grassmann, Pauli, Dirac: special relativity in the schoolroom. In H. J. Petsche, A. Lewis, J. Liesen, and S. Russ, editors, *From Past to Future: Grassmann's Work in Context*. Birkhäuser, 2011.

[44] Martin E. Horn. Loesung einer Aufgabe zu linearen Gleichungssystemen aus der Han-Dynastie mit Gaalop als Taschenrechner-Ersatz. In *Beitraege zum Mathematikunterricht BzMU*, 2017.

[45] J. Hrdina, A. Navrat, and P. Vasik. 3–link robotic snake control based on cga. *Adv.Appl. Clifford Algebr.*, 26(3):1069–1080, 2016.

[46] J. Hrdina, A. Navrat, P. Vasik, and R. Matousek. CGA-based robotic snake control. *Adv.Appl. Clifford Algebr.*, 27(1):621632, 2017.

[47] J. Hrdina and P. Vasik. Notes on differential kinematics in conformal geometric algebra approach. In *Advances in Intelligent Systems and Computing*, volume 378, pages 363–374. Springer-Verlag, 2015.

[48] Kenichi Kanatani. *Understanding Geometric Algebra*. Taylor & Francis Group, 2015.

[49] Ulrich Kortenkamp. *Foundations of Dynamic Geometry*. PhD thesis, ETH Zuerich, 1999.

[50] Hongbo Li, David Hestenes, and Alyn Rockwood. Generalized homogeneous coordinates for computational geometry. In G. Sommer, editor, *Geometric Computing with Clifford Algebra*, pages 27–59. Springer, 2001.

[51] Dierck-Ekkehard Liebscher. *Relativitaetstheorie mit Zirkel und Lineal.* Akademie Verlag, Berlin, 1991.

[52] P. Liljeback, K. Y. Pettersen, O. Stavdahl, and J. T. Gravdahl. *Snake Robots, Modelling, Mechatronics and Control.* Springer-Verlag, 2013.

[53] Maxima Development Team. Maxima, a computer algebra system. version 5.18.1. Available at http://maxima.sourceforge.net/, 2017.

[54] R. M. Murray, L. Zexiang, and S. S. Sastry. *A Mathematical Introduction to Robotic Manipulation.* CRC Press, 1994.

[55] J. Ostrowski. *The Mechanics of Control of Undulatory Robotic Locomotion.* PhD thesis, CIT, 1995.

[56] Josep M. Parra Serra. Clifford algebra and the didactics of mathematics. *Advances in Applied Clifford Algebras*, 19:819 – 834, 2009.

[57] Christian Perwass. *Geometric Algebra with Applications in Engineering.* Springer, 2009.

[58] Christian Perwass. The CLU home page. Available at http://www.clucalc.info, 2010.

[59] Christian Perwass and Dietmar Hildenbrand. Aspects of geometric algebra in Euclidean, projective and conformal space. Technical report, University of Kiel, 2004.

[60] Venzo de Sabbata and Bidyut Kumar Datta. *Geometric Algebra and Applications to Physics.* Taylor & Francis Group, 2007.

[61] Steve Sangwine. Colour image edge detector based on quaternion convolution. *Electronics Letters*, 34(10):969–971, 2015.

[62] J. M. Selig. *Geometric Fundamentals of Robotics.* Springer-Verlag, 2004.

[63] Florian Seybold. Gaalet – a C++ expression template library for implementing geometric algebra. Technical report, 2010.

[64] Garret Sobczyk. David Hestenes: The early years. *Foundations of Physics*, 23:1291 –1293, 1993.

[65] G. Soria-Garcia, G. Altamirano-Gomez, S. Ortega-Cisneros, and Eduardo Bayro Corrochano. Fpga implementation of a geometric voting scheme for the extraction of geometric entities from images. In *Advances in Applied Clifford Algebras Journal*, Sept. 2016.

[66] Christian Steinmetz. Examination of new geometric algebras including a visualization and integration in a geometric algebra compiler. Master's thesis, 2013.

[67] Florian Stock, Dietmar Hildenbrand, and Andreas Koch. Fpga-accelerated color edge detection using a geometric-algebra-to-verilog compiler. In *Symposium on System on Chip (SoC)*, 2013.

[68] Richard Wareham. *Computer Graphics using Conformal Geometric Algebra*. PhD thesis, University of Cambridge, 2006.

[69] J. Zamora-Esquivel and E Bayro-Corrochano. Kinematics and differential kinematics of binocular robot heads. In *Proceedings 2006 IEEE International Conference on Robotics and Automation.*

Index